Cinema

Concept and Practice

Cinema

Concept and Practice

EDWARD DMYTRYK

FOCAL PRESS
Boston   London

In every age theory has caused men to like much
that was bad and reject much that was good.
—Aldous Huxley, *Collected Essays*

Library of Congress Cataloging-in-Publication Data
Dmytryk, Edward.
   Cinema : concept and practice / Edward Dmytryk.
     p.   cm.
   "Filmography of Edward Dmytryk" : p.
   ISBN 0-240-80002-8
   1. Motion pictures—Philosophy. I. Title.
PN1995.D54   1989
791.43—dc19   88-11685

British Library Cataloguing-in-Publication Data
Dmytryk, Edward
   Cinema : concept and practice.
   1. Cinematography
   I. Title
778.5'3
ISBN 0-240-80002-8

Butterworth Publishers
80 Montvale Avenue
Stoneham, MA 02180

10 9 8 7 6 5 4 3 2 1

Printed in the United States of America

Contents

Introduction

What is "Film"? Experts are in agreement that, at its best, film-making is an art, and we are immediately in the soup. Because the next question is, "What is Art?" And that question has stumped truth-seekers since the first primitive man or woman traced the outline of a wooly mammoth on the walls of a firelit cave.

This particular inquiry began after I retired from active filmmaking, started to teach, and found that I didn't really know what I had been doing for more than sixty years. I had just been doing it. The discovery was traumatic. I decided to go to the professionals—no, not those who *made* films, since in all those sixty-plus years I had never heard even one working director try to define his methods or his product (wasn't it Wagner who said, "When you create, you do not explain"?)—but to those professionals who *study* the filmmakers' work and then tell each other what it was the filmmakers were doing and how they were doing it.

It puzzled me that these aesthetes unconditionally accepted as artists the person who chopped out a totem pole, the sculptor who chiseled out *Moses*, and the innovator who cut off an eight foot section of train rail, stuck it in a block of concrete, and called it a statue, but balked at placing the director of a *Hopalong Cassidy* in the same category as the director who made *Doctor Zhivago*.

I would like to hypothesize, quite simply, that Art is by far the greatest component of a "field," a field that includes all those activities of man which have no direct effect on his everyday life, and the absence of which need not diminish in the least his basic physical existence. "Film" fits such a category to a tee.

However, film theory has demonstrated that categorizing film serves only to give it a certain status; it does little to explain it. A few definitions will make the point.

- The camera [is] a medium particularly equipped to promote the redemption of physical reality—*Siegfried Kracauer*
- Film is a dream—*Dr. Serge Lebovici*
- The true material of the sound film is, of course, the monologue—*Sergei Eisenstein*
- The chase seems to be the final expression of the motion picture medium—*Alfred Hitchcock*
- The cinema is the dynamism of life, of nature and its manifestations, of the crowd and its eddies—*Felix Mesguich*
- Films must be drawings brought to life—*Hermann Warm*
- [Cinema is] the art of the movement and the visual rhythms of life and the imagination—*Germaine Dulac*
- Films cling to the surface of things—*Siegfried Kracauer*
- This music of light has always been and will remain the essence of cinema—*Walter Ruttman*
- Cinema is only a secondary task in the world struggle for revolutionary liberation—*Dziga Vertov*
- The cinema is a system of signs whose semiology corresponds to a possible semiology of the system of signs of reality itself—*Pier Paolo Pasolini*

Gertrude Stein said it better.

There are many such definitions in my files, but these few will serve as a "system of signs" that, like a politician's stump speech, point in every direction. The convenient thing about art is that "you pays your money and you takes your choice."

I was surprised to learn that even the theorists agreed that there is no general theory covering the scope of film. Theories of the special case there are in plenty, and each strips the substance of film to the bare bones of formula and formalism. But I was even more surprised to learn that writers and actors, craftsmen I have always considered indispensable, were mentioned only rarely, if at all, and then usually as "noise" to bother the director. $E = c^2$ hardly makes sense—the m is missing.

Speculation about what film is, or should be, is a useless exercise if one fails to take into account the people who make it. This stipulation immediately removes the auteur theory from serious consideration as, for that matter, it does most others. As a commercial enterprise or a work of art, film cannot be compared to a painting, which is usually the work of a single artist, or even to a play, which is not. A film more closely resembles a symphony—which also has a director.

One such, Toscanini, was a great artist, but he was no auteur. A special interpretation of the score was his, but the score itself was not; a hundred other artists, the members of the orchestra, contributed to the realization of his concept. If some of these were less than A-one, so was his interpretation.

The same is true of film. On any particular motion picture the limit of the filmmaker's achievement is a function of the combined contributions of the people he works with, and a viable theory of film must take them all into account.

Certainly, two of the most important contributors are the writer and the actor. I cannot think of a single film genre that does not require at least a thread on which to hang its scenes—a thread supplied by a writer—and only abstract animation can exclude the actor. So it disturbed me to realize that through all those heedless years, forty of them as a director, I had been deaf to all that "noise," and I decided to work out my problems with pen, paper, and typewriter in an effort to discover whether *any* theory of narrative filmmaking was worth a plugged nickel.

The author and his crew on location for **Alvarez Kelly**. *A successful film is a collaborative effort; sharing the load makes the difficult times easier. Photograph courtesy of Columbia Pictures.*

1

The Collective Noun

The more I see of the performing arts the more firmly I am convinced that a precise definition of the ideal "Film" would differentiate it from any of its cousins. With the sole exception of film, the performing arts are appropriately named; in all instances the visible artists in this branch of the arts *are* performers. Whether they stand before an audience or a television camera to deliver one-liners, "read" a fine author's golden words, dance a noted choreographer's pas de deux, or sing an operatic aria or one of tin-pan alley's clever ditties, they are *performing*. And in these areas a fine performance can be a gratifying experience. The viewer watches, and listens, and usually appreciates the performer more wholeheartedly than he does those who created the material performed.

But in films a "performance," so recognized, rings falsely.* As we have already seen, most definitions of film include the words "life" or "reality." Regardless of whether the filmic technique stems from Eisenstein or Godard or Hollywood, the actors on the screen must "be" the human beings the viewers watch, human beings who, if the film is skillfully made, soon become familiars, and the viewers' emotional response should increase along with their involvement

---

*It is unfortunate that film's founding fathers little realized their plaything's potential, and borrowed their descriptive vocabulary from its antecedents. Because of this we have no adequate word for "being" on the screen, and the word "performance" must of necessity be used in any discussion of the screen actor's art. Reluctantly, it will be so used in this book.

with those who live on the screen, whose joys and sorrows they feel as their own, whose existence they pity or envy or dream of sharing. Only after the vicarious experience of being part of such lives begins to dissolve into reality will analysis and appreciation of the film's separate elements, including the art of the actors and the skills of the technicians who made the film, begin to encroach upon the viewers' attention. At least, that's how it should be.

Alas, how often are things the way they ought to be? Ninety-five percent of all film made does not even come close to meeting the standards of an ideal motion picture. But to avoid any possible misunderstanding, let us first exclude all filmmaking which lies outside the scope of this discussion. There are a number of fields that lend themselves to filming; a rather extensive one covers industrial subjects and techniques in their different phases, from explication of manufacturing processes to the selling of the products to agents and dealers. For the consumers of those products and nearly everything else, there are the ubiquitous commercials, without which neither our economy nor our society could now survive. Rock video, designed to sell or display "pop" music, and styled to interest and excite the youthful fans, is a fast-growing newcomer in the television and home VCR fields.

There are the documentaries, which inform the viewer on the states and activities of society and the environment, and docudramas, which do the same thing with less fact and more fiction, and sometimes do it better. There are children's cartoons, full length narrative cartoons, puppet films, educational films, "art" or experimental films which range through all the genres known to painting, and a few others, such as film verité, which the more static media cannot accommodate.

Then there is the self-defining "narrative" film, a genre which has sired a large family of overlapping subgenres, but which I will separate roughly and arbitrarily into "theatrical" and "cinematic" films.

Of all the foregoing classifications, excluding only commercials, the public is acquainted primarily with the documentary and the narrative film. Today, the documentary belongs exclusively to television. Although documentarians like Flaherty, Schoedsack, Lorenz, and others have been acknowledged as superior artists whose work attracted sizable audiences, the narrative film unquestionably overshadows documentaries and all the other classifications by a very wide margin. It is primarily this "Film," whether shot on film, tape, or a combination of both, and shown on a television set or on a theater screen, which will be discussed in the following pages.

Sixty-five years ago, Jacques Feyder, a French film director, wrote, "Everything can be transferred to the screen, everything expressed through the image. It is possible to adapt an engaging and humane film from the tenth chapter of Montesquieu's *L'Esprit des Lois*, as well as Nietzsche's *Zoroaster*." This somewhat optimistic analysis was written in the days of the "silents," but now that the screen speaks it is both much easier and much more difficult to realize Feyder's dream.

It is easier because the marriage of sound and image makes it possible to achieve a certain depth of meaning even when dealing with a difficult concept. When dramatizing a complex human emotion, condition, or conflict, words are often too specialized or ambiguous and the rhetoric too literary for the person of average education and understanding.* And isn't it this person we should be most concerned to reach?

The versatility of film permits the use of a relatively straightforward verbal language while the accompanying images serve (broadly speaking) to "diagram" the scene's intellectual or dramatic richness, and to eliminate sources of ambivalence, thus making it accessible to nearly all levels of understanding.

This aspect of film was brought home to me quite dramatically a few years ago at the California Institute of Technology. Following an annual custom, on Alumni Day a number of scientific lectures were presented on campus, all with the aid of specially made films. The combination of nontechnical language and filmed demonstrations clarified some very abstract scientific concepts and activities for audiences that contained many persons with no scientific background whatever. Not exactly new, but a great advance over the magic-lantern slide show.

Closer to home, a number of successful narrative films support this point of view. The film *Amadeus*, beyond its excellent dramatization of character and situation, beyond its presentation of eighteenth century customs, costumes, and behavior, brought the beauty of Mozart's music and more than a glimpse of some of the technical

---

*See Allan Bloom, *The Closing of the American Mind* (New York: Simon & Schuster, 1987).

aspects of his art to millions of viewers, most of them with little knowledge of, or previous interest in, music at this level. The film even impressed a respectable number of those who consider anything beyond punk rock a total waste of time.

The truth is that this happy combination of words, images, and sounds in the interest of deeper content is not often realized, but that is due to the scanty supply of talented filmmakers rather than to the medium's inability to deliver. As has frequently been pointed out in the world of computers, theorists excepted, a viewer gets no more out of a medium than that medium's manipulator puts into it.

However, Feyder's dream has also become more difficult of realization because words have once more largely taken the place of effective imagery. Narrative film has been engaged in a tug-of-war with the theater that the theater, through an almost "Chinese" process of absorption, seems to be winning. Although purists continue to maintain that the cinema should be *only* a medium for images, most of today's films are no more than richly illustrated plays. Any argument on this point has been made moot by the reality. And though I will try to make a case for the modified cinematic film as the truest and best art of the screen, only the tunnel-minded will rule out any narrative form which adapts itself to film and succeeds in interesting and entertaining the viewer.

At this point there must be an unambiguous understanding of the word *entertainment*. The expression is equivocal, but in theatrical terms most people think of it as "something which amuses." I will always use it in the following senses of the word.

1. *Entertainment:* That which affords interest and amusement.
2. *To entertain:* To engage, keep occupied the attention, thoughts, or time of a person.*

In other words, entertainment is not only *Beverly Hills Cop* or *Star Wars*, it is also *Amadeus, Ghandi, Missing,* and *One Flew Over the Cuckoo's Nest.* A film may amuse, inform, explain, analyze, or deliver a message; as long as it holds the viewer's attention, it is entertaining.

Why all the emphasis on entertainment, or interesting the viewer?

Oxford English Dictionary.

Tom Hulce as Wolfgang Amadeus Mozart in **Amadeus**. *The actor's performance is obviously a crucial part of a film, but for the film to be successful, the other parts, including the sound and narrative imagery, must also be excellent. Photograph courtesy of Orion Pictures.*

Why not, like some purists, consider only the artistic excellence of the film and the skill of the filmmaker? Because without the viewer no film lives—it is merely a long succession of photographs, carrying no meaning and no emotion. Its significance can be brought to life only through the empathetic reactions of a viewer's involvement. And even the diehard purist must admit that film is a business as well as an art. Building a motion picture requires the services of a large, highly-paid professional crew, the contributions of an exorbitantly-priced group of "artists," the utilization of tons of expensive equipment, and time.

"Highly-paid," "exorbitantly-priced," and "expensive" equals money—a great deal of money. Today, the cost of a very modest film would keep an American family in comfort for a lifetime. The financiers who supply such money, whether through the studios or though individual producers, would like to retrieve their investment along with as much profit as their money would earn while sitting in a bank. The dream, of course, is for a great deal more.

The source of the recouped investment is the viewer. If he does not find the film to his liking there is little recoupment and no profit at all. Such is the fate of many films—in fact, of most films—but for the producing studio, disaster is extremely rare. Ancillary revenues from sources such as video, television, toys, and music albums, all fatten the kitty, and unless it goes as far over the edge as United Artists' *Heaven's Gate,* a studio can "average out." One hit will make up for a dozen flops, and the losers will at worst furnish a useful tax loss for the parent conglomerate. On the other hand, the independent who gambles it all on one film is challenging one of the world's highest risk businesses. His chance for making a killing is small indeed. But, fortunately for the good filmmaker, there is never a shortage of gamblers, and the independents, as well as the studios, manage to struggle along, their hopes kept new-penny bright by the few who succeed.

Profit may be the main goal of the producing entity, but most filmmakers dream of creative freedom, and the quickest, perhaps the only way to get it is to make films which attract large audiences. With so much money riding on each production, the director with a record of box-office success is obviously in demand. The greater the success the greater the demand, and the greater the demand the greater the amount of freedom requested, and received, by the film-

A successful motion picture always benefits from a well-known cast. For **Broken Lance**, *this investment paid off—the film won an award for the best Western of 1954. On this lunch break near Nogales, Arizona, the author is surrounded (on his right) by Spencer Tracy and (on his left) by Robert Wagner and Earl Holliman. Across the table are Jean Peters, Richard Widmark, and, with his back to the camera, Hugh O'Brien.*

maker. It follows that the filmmaker with a poor success rating soon loses the opportunity to make demands—or films.

Such are the film facts of life. These simple and quite logical parameters pose a number of problems and offer a few opportunities. Many of the films which rack up huge grosses are, to put it politely, tripe. But if the tripe sells, those who produce it get greater freedom to turn out more tripe. The bright side of the picture is that each year a few films, made by thoughtful and talented filmmakers, also manage to attract large film audiences, audiences composed of adults who rarely visit movie houses, and the more responsive members of the youthful community. Together, these two groups can make up an impressive profit-turning array of spectators.

It must be obvious that the filmmaker who can produce a quality film of substance which can also attract and hold the attention of a mass audience must possess abilities far transcending the ordinary. His or her message must be delivered in a manner that is understandable to the average person, yet deep enough to please those who demand nourishment in their entertainment diet; the film must not be pompous, pretentious, pedantic, or overbearing—and it *must* entertain. It can be as raucous as a Marx Brothers comedy, which the discriminating will easily recognize as a sharp satire of our social and political structures; as brilliant as *Doctor Zhivago*, which deals with some of the greatest upheavals of revolutionary Russia while holding the viewer's complete attention with a superior love story; or as stark and stomach-turning as *One Flew Over the Cuckoo's Nest*, which lays bare the tragedy and the cruelty of a mental institution while supplying more laughs than most Chevy Chase comedies. Each of these furnishes amusing or emotional entertainment to its viewers while providing substance for those thirsty enough to require it and intelligent enough to demand it. And they all bring our attention back to the filmmaker—which is really a collective noun.

Fans have always been the lifeblood of film. The photograph depicts Grauman's Chinese Theatre, 1930, during the world premier of Morocco, starring Marlene Dietrich. Photograph courtesy of Bruce Torrence Historical Collection.

2

The Indispensable Viewer

Excluding for the moment the cinematographer, the film editor, and a few other film artists, the chief contributors to the substance of any film are the writer, the director, the actors, and the *viewers*. Outside of their indispensable involvement in the business aspect of films, the viewers' participation is indirect, but it is of extreme importance. Although they have no specific input into the production, they force the writer, the director, and the actors to make audience-oriented decisions at nearly every step of the way—decisions which are not financially inspired, but have to do with purpose and incentive.

For many filmmakers, of course, the incentive is money, but though it is true that a few artists get rich, I have never known a *good* artist who made wealth a prime goal in life. The fact that artists get paid well if they do well what they love doing is beside the point. What drives most filmmakers is that they are preachers of a sort, with enough ego and arrogance to assume, as Tolstoy pointed out, that what they create is worthy of the attention of others. Since they are, generally speaking, quite human, they also need acceptance, and only an audience can give them that. It takes sheer stupidity or, more probably, sour grapes to maintain that one need not be concerned with the viewers' dreams and desires, that, willy-nilly, the viewers must accept what they are given. If the often-mentioned creative vision cannot spark a response in the viewers, it cannot sustain a life of its own. Would the *Mona Lisa* be *the Mona Lisa* if it did not stimulate a response in millions of minds? And if Beethoven's Fifth did not excite millions of hearts would it still be a great symphony? In painting or music the response can start slowly and build; it can even, in time, overcome opening failure. But in films, the response must be strong and immediate. Rarely does an unpopular effort receive a second chance. The creative vision is capricious; it needs the energy of the viewers to give it body and lasting life. Too many self-proclaimed "creators" have learned the bitter truth; unless something about your work intrigues him, the viewer will turn his back on it and you.

What *are* the film viewer's dreams and desires? Why will he spend money that might better feed him, to buy his way into a motion picture theater? Many surveys have been conducted over the years; one of these, which suits my thesis best, was made by Wolfgang Wilhelm.

- A housewife said, "The film is more life than the theater. In the theater I watch a work of art which appears to be elaborated. After a film performance I feel as if I had been in the middle of life."
- Another woman said it more poetically, "In the cinema I dissolve into all things and beings."
- A nurse responded, "A good film helps me to get in touch with people and with life."
- And a businessman said, "The less interesting the people I know the more frequently I go to the movies."
- A young writer finds it a substitute, "One would like to get something out of life, after all."
- And a student confessed, "Some days a sort of hunger for people drives me into the cinema."

From these few comments alone one could derive a set of sound guidelines; rules that would help ensure a viewer-oriented film, yet in no way inhibit the filmmaker in the pursuit of his creative vision. The last comment is perhaps the most important; it states flatly what the others hint at. Today, more than ever, many people from all walks of life suffer from loneliness, from alienation.* A good film can give them the illusion of partaking of life in its fullness. The people on the screen are their friends who take them into physical and emotional worlds they may never otherwise experience. Yet one of our current problems is that far too many films fulfill the viewers' imagined physical requirements to the point of satiation but fail completely to address themselves to their emotional needs.

Since the beginning, many filmmakers and film theorists have held that people look for an escape from their daily tribulations through films. But statistics on the sale of self-improvement books and the popularity of self-improvement preachers clearly point to the fact that, rather than looking for a way to escape life, most viewers want to learn how to *live* it. Philosophers and psychologists tell us ever more urgently that alienation is one of our more critical social problems—alienation from family, from friends, from church, from society. In many ways the world is more fragmented than at any time in history. There are no longer any monolithic religions, or homogeneous nationalities. People within nations have been polarized into political parties, factions, sects, splinter groups, and so many major and minor "isms" that only the extreme doctrinaire at either end of the political spectrum is sure he knows exactly what patriotism is.

As they drift away from moral and spiritual revelation (and the drift is inevitable in our scientifically oriented society), many find themselves viewing the old standards and principles through the wrong end of an out-of-focus spyglass, and they look for a new and brighter "vision" before vertigo and nausea set in. To the clear-headed and well organized filmmaker this is all grist for the mill, an unparalleled opportunity for making films of substance. Problems of alienation can be isolated and analyzed; a variety of solutions, each depending on the filmmaker's point of view, can be advanced; some of them may help the viewer to a better understanding of himself and his place in a confusing world. It is not enough to point out the problems; news broadcasts do that with a frightening tenacity. Healing points of view are vital. The Marx Brothers could rip society apart, but someone must follow to sweep up the pieces and recombine them to a new and better purpose.

---

*It is my firm belief that if the motion picture theater survives it will not be because of its superiority to television, but because of the loneliness factor.

Learned explanations and clarifications customarily serve only to confuse the average citizen. On the other hand, film is without equal in its capacity to simplify without being simplistic. Ineptly used, that capacity is a weakness which many intellectuals attack with ferocity, and often with justification, but the intelligently made film is the ideal medium for bringing even the most complex human problems into the area of normal understanding.

To do so it makes use of its most distinctive attribute—the ability to "show." It is axiomatic that no one ever says, "I believe everything I read," or "I believe everything I hear." But all languages are replete with phrases that testify to the worth of the image. For example, "One picture is worth a thousand words" or "*Show* me, I'm from Missouri." The viewer is aware that the filmmaker can practice all sorts of legerdemain with trick shots and double exposures, for example, but if he is given no cause to assume that the filmmaker wishes to deceive him, he will usually accept what he sees—*if* it is well done and entertaining.

A short segment of *The Reluctant Saint* demonstrates simplicity of approach and simplicity of explanation. Based on a real character, the story takes place in the middle of the seventeenth century. Joseph (Maximilian Schell), a simple-minded young layman at his uncle's monastery, is roasting chestnuts outside his domain, the monastery barnyard. It is late in the evening and a visiting dignitary, Bishop Sturzo (Akim Tamiroff), escaping from the over-attentive environment of the monks' quarters, is attracted to Joseph's fire. He sits down for a chestnut and a chat, and in a quiet moment he gazes up at the stars. (The excerpt is written in a master shot for easier reading.)

BISHOP STURZO
Beautiful night. Look at those stars!

JOSEPH
(looks up—after a beat)
Have you ever noticed—the longer you look,
the more stars come out?

BISHOP STURZO
(likes the observation—but testing)
I wonder why?

Joseph wrinkles his brow—shrugs. He has passed the
test.

                    BISHOP STURZO (cont.)
          I'm sure Brother Orlando would have an
          explanation.

                         JOSEPH
                    (sincerely)
          Oh, yes! Brother Orlando knows everything.

                    BISHOP STURZO
                    (sourly)
          I just heard him trying to explain the
          Trinity. . . .
                    (a hopeless shrug)
          Now it's more of a mystery than ever.
                    (he shakes his head)
          You know, Joseph, it always puzzled me—the
          Trinity. Never could understand it—just took
          it on faith. What about you? Does it trouble
          you?

Joseph looks a little shocked. He considers the matter.
Then he holds up three fingers.

                         JOSEPH
          Three persons in one God—
                    (raises one finger at a time)
          Father, Son, and the Holy Ghost.

                    BISHOP STURZO
                    (smiles)
          Yes, but that's for a child.
          It's small comfort for a bishop who's
          supposed to know theology.
                    (he leans in confidentially)
          Look, Joseph, I'm a peasant—just like you. A
          practical man. I understand what I feel—
          what I see.

Joseph considers the idea, then stands up and takes
down a blanket which hangs on a peg outside a stall.
He smiles at the bishop, holds up the opened blanket.

JOSEPH
You see? One blanket. . . .

Then, he folds the blanket in three pleats while
holding it up off the ground much as a housewife
might fold a sheet, and counts:

JOSEPH (cont.)
One . . . two . . . three. Three folds in one
blanket. Three persons in one God.

The bishop is delighted with this simple but direct
mind.

BISHOP STURZO
Brilliant, Joseph! Simply brilliant!

END OF SEGMENT:*

The explanation really explains nothing, but it does a better job
of illustrating this abstract concept than any theological argument
I have ever heard or read. It also demonstrates the superiority of
"show" over "tell."

---

*From *The Reluctant Saint*, Columbia Studios, 1962. Script by John Fante and
Joseph Petracca.

Great comic sequences, as much as any dramatic scene, foster a sense of reality within the viewer. The situation depicted in this photograph, from Harold Lloyd's Safety Last, *is funny partly because it is believable. Photograph courtesy of Harold Lloyd Pictures.*

3

You'd Better Believe It

Believability! Even a cursory analysis of scholars' definitions and viewers' dreams and desires clearly indicates that *believability* is the narrative film's most essential ingredient, and its technical aspects, no matter how artistic or professional, are of value only if they serve to sustain it. When viewers buy their tickets, they are in a very real sense putting their trust on the line, and if directors wish to draw them into their "visions" and keep them there, they must respect and sustain that trust. If the film betrays belief, even for a brief moment, the viewer's trust is weakened; an implicit compact has been violated.

The preceding statement will be acknowledged as admissible for serious drama, yet its validity is most graphically demonstrated in broad comedy. Buster Keaton, who understood his genre far better than most theorists, once said that a comic situation should never appear to be ridiculous. An examination of any outstanding comedy sequence will substantiate his point of view.

In a classic scene from *Safety Last*, Harold Lloyd hangs by his fingertips from the hands of a tower clock several stories above a city street. Although the viewers are deeply concerned with Lloyd's predicament, they laugh at the comedian's antics as he struggles to save himself. The believability of the situation increases the viewers' concern and enhances their enjoyment. In *The Gold Rush*, Charlie Chaplin and Mack Swain are stranded in a one-room cabin by a raging blizzard which keeps them snowbound for many days. Eventually, extreme hunger drives Chaplin to boil and eat his shoes, while Swain reacts to his own hunger by hallucinating.

The sad fact that every generation has its records of people driven by starvation to eat almost anything, including other people, makes the situation believable. But what makes this innately tragic scene extremely funny is that Chaplin eats the boiled boot delicately, as if he were a gourmet dining on pheasant under glass, while the hallucinating Swain, a huge hulk, sees and stalks Chaplin as if the comedian were a man-sized chicken. If the basic predicament were not so completely credible and worthy of the viewer's concern, the comedy would be seen as completely contrived.

There are a number of situations, however, where believability can bring us to the brink of disaster. In the present context, the foregoing examples are the reverse of what happens in more serious films. Comedy scenes of suspense carry built-in relief; laughter is their reason for being. In straight drama, suspense, or good horror films, that is obviously not the rule. When such films are realized, scenes of suspense can reach a point where viewers react in fear— a few may scream—even though something at the back of their minds assures them this is only make-believe and, whatever happens, the endangered character(s) cannot actually be hurt.

Such a feeling will remain subliminal until the suspense becomes unendurable, at which point, like a dreamer rejecting a terrifying nightmare, viewers will pull themselves back into their *real* reality—the theater and the people around them. That is why "safety valves" are essential. The more compelling the scene the greater the obligation to provide viewers an opportunity to extricate themselves from intolerable emotions, such as fear, suspense, and sorrow, *before* they are forced to deny the film's reality.

But I'm ahead of myself, slipping into an analysis based on a filmmaker's point of view rather than that of a film theorist, speaking of emotion and involvement rather than of how these two factors can best be realized in a medium composed of two-dimensional images accompanied by mechanically recorded sound, each in turn much bigger and louder than life.

Two essential elements of a script, plot construction and character development, are much the same in all of the narrative media. But when the end product is meant for the screen, the unique tools and techniques of the motion picture make possible special treatment

Charlie Chaplin was a master of the comic film. In this scene from The Gold Rush, *he and Mack Swain fight off hunger. Photograph courtesy of Charles Chaplin Films.*

of plot and character which can render that product more believable and more effective in *reaching and touching the viewer*.

To be sure, the reasons for making commercial films have little to do with the tools of the trade; content, if considered separately, is more important than structure. Certainly, technique without substance is purely of parochial interest. On the other hand, effective and challenging ideas are, in themselves, not enough. By a creative alignment of carefully chosen words an exceptional author can elicit more emotion and illuminate more ideas in a single paragraph than a routine scribbler can manage in a whole book. Filmmakers must have an equal talent; the radiant "visions" they so frequently invoke become tiresome twaddle if transferred to the screen in unknowing fashion.

Some mastery of filmmaking techniques is essential for effective translation of vision into image, but in the working world it is at this level that misunderstanding persists, and that much of the conflict between writer and director arises. As a rule, writers are *not* trained to think in terms of cinematic images, action, and reaction; words are their stock in trade. But film is not a literary medium. A good director thinks more of images, metaphor, and reaction than he does of words and, assuming equal understanding of the content, the filmmaker with the greater knowledge of the creative use of film techniques is more likely to enrich the given values of the film's substance and to enhance its appeal to viewers.

Which gets us back to syntax and style. Unlike the birthing of original ideas, the basic rules of film technique can be taught and learned, though creative manipulation of those rules into an individual style is, alas, a matter of talent. But before syntax and style can be considered, even at an elementary level, the essential "words" of film grammar, the *shots* or *angles* which are the building blocks of film construction, must be fully understood.

The choice of a *set-up*, as the shot or angle is more commonly called on the set, is no mystery to an experienced filmmaker, and he rarely thinks of it in aesthetic terms. For him the set-up is determined by the substance of the scene. It is as simple as that. It may be an advantage if the shot is also beautiful, but more often than not beauty befogs the essence of reality, and the primary purpose of a particular set-up is to offer the viewer the best possible presentation of the scene or, most often, some particular part of it. Ignoring for the present the personal embellishments or distortions the director may supply, it follows that the dramatic demands of different scenes, or portions thereof, require set-ups that are best suited to those demands, and here practice will often depart from theory, since the "reading" of such demands will vary with the "reader."

In film's crawling stage the *long shot* was the *only* shot in the filmmaker's repertoire and, out of respect for age, it is proper to start with its analysis even though, today, it is probably the least common set-up on the director's list. It is impossible to define any shot exactly in terms of size, and the long shot is the prime example of this difficulty. It includes a near infinite variety of shots, ranging from little more than a full figure (in height, not amplitude), through a bird's eye view of, say, Death Valley, to a shot of the earth taken from an orbiting satellite.

The development of a battery of closer shots in what must have seemed to be a contradictory search for dramatic truth destroyed the all-purpose aspect of the long shot, but its supporters gave ground grudgingly. Long after Griffith (or Billy Bitzer) introduced the *close shot*, important critics objected to the sight of human bodies apparently sliced off at the waist and with no visible connection to an imaginary ground below the bottom of the screen. Indeed, for many years it remained compulsory to begin each new sequence with a *full* or *establishing shot* (two of the long shot's alter egos) whose purpose, as the term indicates, was to establish the setting or location of the sequence and the relative positions of the actors in it.

To be sure, techniques have changed with time. But the films of John Ford and David Lean still exemplify the proper use of the exterior long shot more eloquently than those of any other modern filmmaker. Their long shots are not only compositional masterpieces, they do what *every* shot in a film should do: they further the establishment and the development of the film's characters. An Oriental artist might symbolize "winter" with a bare bough, a bird ruffling its feathers against the cold, and a few artfully designed snowflakes. Under Lean's direction, "winter" becomes an entity, a motivating force in the film. In *Doctor Zhivago*, vast areas, buried under a claustrophobic white shroud, inform the viewer of the Russian winter's influence on the character and the attitudes of the land's inhabitants far more effectively and believably than any aesthetic presentation of the idealized winter could possibly do. And every viewer will sense, if only subliminally, that a dweller in Ford's

The "white claustrophobia" of winter is a constant presence in David Lean's Dr. Zhivago. *Photograph courtesy of M.G.M. Enterprises.*

impressive but stark Monument Valley must have a different conception of the world and its people than, say, an urbanite born and reared in Philadelphia.

The Western painter, Charles M. Russell, pioneered a style that is especially compatible with outdoor films. His magnificent depictions of the plains, mountains, and mesas of the Northwest never feature the landscape alone; hunters, cowboys, and Indians, often on horseback, and usually in violent movement, dominate the foreground and lend vitality to his compositions, but the background is always there to awe and inspire the viewer. On film, the proper balance of the two elements is a function of time and cinematic instinct, since too much attention given to one will often detract from the other. The beauty of the setting should attract the viewer's attention, but it should be admired in context, not as an independent composition. It should not interrupt the flow of the story or diminish, even momentarily, the importance of the characters in the shot. When a narrative film becomes a travelogue it also becomes a bore.

The more constricted and more commonly used *interior* long shot discloses or establishes a location such as an *indoor* sports arena, the interior of a cathedral, or a convention hall. The approach to the most effective use of this shot, indoors or out, is best described in an excerpt from *On Filmmaking**:

> If a director is filming a football game, he might choose to start his sequence with a long shot of the stadium. This would serve to "establish" that there is a football game in progress with a large crowd in attendance, nothing more. Now, think of the sequence as starting in a different way: a close shot of the quarterback barking signals; then, in a series of cuts, the center hands him the ball, the play develops, the receiver streaks down the field, the quarterback evades a tackler and releases the ball, the receiver catches it in the end zone. TOUCHDOWN! Now, for the first time, a full shot of the stadium as the crowd rises to its feet in a screaming reaction. Here, the long shot is employed for dramatic emphasis as well as an establishing shot. There is no confusion, for even though, in the earlier set-ups the background has been exposed only incidentally, at no time would a viewer be in doubt that the game is being played on the gridiron.*

*From Edward Dmytryk, *On Filmmaking* (Stoneham, MA: Focal Press, 1986), p. 233.

The introduction of closer shots eliminated the presence of the "fourth wall"—the audience—and encouraged far more flexible staging. It also mandated the development of more advanced editing techniques. But for many years cutting from a long shot to a *close-up* was taboo. It was deemed necessary to approach the closer angle through one or more intermediate steps, the number being determined by the relative sizes of the shots at either end. In other words, the long shot was followed by a *medium shot*, then probably by a close shot before the desired close-up (there is a difference) could be used, a routine that involved careful planning. The development of a high degree of viewer sophistication eventually rendered both the practice and the medium shot obsolete. Now the screenwriter who types *medium shot* is using a convention which long ago ceased to have any exact meaning. He is also admitting he has no specific set-up in mind.

It is a truism that development of new equipment fosters new techniques which, on occasion, lead to improvement in an art or a craft. Perhaps the best example of a creative set-up brought into existence by the introduction of a new tool is the modern *master shot*, which has become one of the two most useful shots in the director's arsenal. This much misunderstood set-up was originally a full shot which, although used in conjunction with closer angles, carried the main burden of the scene since the editor returned to it frequently in order to reestablish the scene's setting or the "real time" of the action. The descriptive term *master shot* was applied to the full shot alone, not to the series of shots with which it was intercut. A simple *opening* or *establishing shot* was never considered a master shot, unless it continued into the scene's mainstream.

The development of the *crab dolly* made the modern master shot possible. Although there are many pieces of special equipment available, in general, directors find the crab dolly the most useful. It needs no tracks, is capable of moving easily in any direction, and is much more maneuverable than a tracked dolly or crane. It can incorporate a series of angles within a single set-up, limited only by the ingenuity of the director and the dexterity of the camera crew, especially the master grip. To put it simply, the camera crew can move the dolly *in* or *out*, from a full shot to a close-up, and back. The camera can be moved *right* or *left* or *obliquely*, and it can be moved *up* or *down* to accommodate limited vertical movement and compositional adjustment. When combined with the actor's freedom to move into or away from the camera, such a set-up, precisely staged and paced, can encompass an entire sequence without recourse to the editor's input. However, if the shot runs too long its precision stifles spontaneity of movement as well as spontaneity of performance. The result is a stagy, awkwardly-timed and noncinematic scene which only a clever editor might salvage, if he has been given the necessary protection shots. There are definite limits to the use of the technique, whose pitfalls are most explicitly demonstrated in Hitchcock's film, *The Rope*. Being a wise man, Hitchcock never repeated the experiment.

That area of set-ups vacated by the general-purpose medium shot has been filled by an assortment of angles whose terms are self-descriptive. This was probably the result of the growing awareness of the film's separation from the theater, and the realization that the opportunity to present more facets, or points of view, increased the possibilities of presenting more and finer shades of screen reality. In effect, the medium shot had said only that we were moving in closer to the characters in the scene, but usually from the same audience point of view. The concept that the *viewer* could be allowed to regard the scene from any number of observation points resulted in qualitative changes in filming techniques. Filmmakers discovered that the viewer could instantly and independently be placed at any point in a *full circle*, even raised or lowered in a third dimension, while the actors maintained their positions relative to each other.

The term, medium shot, now became too ambiguous even for motion pictures, and a more specific nomenclature took its place. Each new term defined the image which filled the frame. A *group shot* includes that collection of characters who are active participants in the set-up in question. It can be the only group in the set, or a particular group separated from a larger gathering such as, for instance, a party or a congregation of spectators. Like the long shot, though to a much lesser degree, it can vary in size, and distinctions in its compositions are indicated by terms such as *loose group* or *tight group*. A tight group is usually limited to five or six people who crowd or overlap the edges of the screen.

A shot which contains fewer than four persons always bears its descriptive title. A set-up of three people is a *three shot*. Two people make up a *two shot*, while a "loose" shot of one person is a *single* or an *individual.* Such self-defining terms may need some modifying, since they can vary in size from loose to tight, or from *full* to *waist figures.* When the single becomes tight we are in close shot territory.

Close shots are normally tight compositions of one or two characters, though a third may occasionally squeeze its way into the frame, as when three heads huddle conspiratorially. A close shot is never fuller than a waist figure, since it is at about this size that facial reaction can readily be seen by the viewer. But it is also at about this size that *body language* begins to lose significance (more of this later).

Except for a few unusual situations the profile two shot, that is, an angle on two people, face to face, profiles to camera, is the dullest shot in the inventory. The characters are too close for meaningful body language and the faces in profile make reading of eye reactions difficult, if not impossible. Such a shot says only that two people are facing each other, nothing more. Under ordinary circumstances it is of minimal value, although, if brief, it may occasionally be of use in a series of cuts.

The preferred two shots are those in which one player is favored over the other, if only to a slight degree, and the best of these is the *over shoulder shot* (O.S. shot) which shares with the skillfully executed master shot a reputation as the most useful set-up in filmmaking. The O.S. shot is usually quite tight—in wide screen format the full head can be as large as that in an individual close-up—and, to use a filmmaker's phrase, it is generally three quarters on the favored character, though the degree of "favoring" is a matter of directorial judgment. In any case, both the eyes "on camera" must be clearly visible so that the actor's reactions can deliver a clear message.

In a single, the close shot becomes a close-up when the bottom line of the frame cuts the subject at upper chest level. From here it can vary in size from "bust" through "choker" (a full head) to a shot of the eyes, or mouth, alone, or even the single eye favored by many horror film makers. The close-up is unquestionably the most abused set-up in films. As Stefan Sharff says, "The shot having dramatic emphasis might be medium, reverse angle, or moving camera shot, rather than the strongest gun in the film arsenal, the close-up."*

*From Stefan Sharff, *The Elements of Cinema* (New York: Columbia University Press, 1982).

Exactly! What many modern filmmakers fail to realize is that the close-up, since it is the most potent set-up at their command, should be used sparingly to deliver unusual reactions or thought processes and at dramatic climaxes. And there are far fewer of these in any film than the average director wishes to believe. The undisciplined use of the close-up for minor purposes weakens its value at those times when its impact should be great. It all started with the exclusivity of long shots, but the pendulum's swing has made it necessary to educate filmmakers to use fewer close-ups rather than more of them. And the best way to do that is to learn the effectiveness of the longer shots in their repertoires.

Incidentally, when a number of people are in the picture, from long shots down to the occasional two shot, it is good practice to fill, or "squeeze" the frame. That is, the shoulders of the characters on the extreme left and right sides of the shot should extend somewhat outside their respective side lines. In effect, this brings the scene closer to the viewers and encourages them to feel that they, too, are part of the group. Players arranged neatly and totally *within* the frame appear to be on a stage (which they are) and the viewers are relegated to the role of spectators. That is an effect the pure filmmaker avoids at all cost.

It will be recognized that the continuing development and increasing use of more revealing set-ups were largely the result of a search for more truth and greater believability. But because filmmakers were often unsure of the viewer's willingness to respond favorably to newer filmic conventions (wrongly so, it turned out), progress was relatively slow (see Chapter 8, page 71). However, it was recognized that the longer shots abetted contrivance and concealment, and soon the camera's encroachment forced fundamental changes in the art of acting (and the interlinked art of editing). The closer the lens the more honest and "real" the actor had to be. Faking was out. Was the crying "dry" or were those real tears in her eyes? Was the smile merely an artificial rictus or a genuine expression of feeling? Was he telling the truth or lying, in his performance as well as in his dialogue?

The first "talkies" exposed the "mechanisms" of a number of Broadway's best-known actors; in close-ups even a child could see the inner wheels go 'round. Many, however, quickly grasped the elemental differences between theatrical and screen acting, and realized that believability was not an artifice, that it was not attained

As this scene from Anzio illustrates, filling or "squeezing" the frame helps bring the viewer into the action of the film. Good filmmakers "keep it tight." This particular shot features Peter Falk. A DiLaurentis Production released by Columbia Pictures.

with a putty nose, by lowering the volume of one's voice, or by simply discarding one's "chest tones," but by creating an aura of truth that completely infuses the artist and inspires "being" rather than impersonation.

The "set-up" is the visual foundation of the film. Here, the author and Dick Powell discuss a set-up for the movie Cornered. Photograph courtesy of R.K.O. Pictures, Inc.

4

The Power of the Set-up

In Chapter 3, set-ups were referred to as the "words" of film. That was a simplification designed to permit elementary definitions and descriptions of the various shots used in filming. In practice such simplification is most uncommon; to continue the metaphor, set-ups are really the sentences, often the paragraphs, on rare occasions even the chapters, that make up the body of any and every film. Without them the modern film would not exist. And without the filmmaker's ability to take advantage of the set-up's responsiveness to personalized manipulation, no matter how crude or how sophisticated, there would be no cinematic style and, of course, no art.

> A set-up is a presentation, a camera point of view of all, or part, of a staged scene, recorded on film, or tape, as a single shot, or "take."

That's a definition. There are no formulas, no rules, for the construction, selection, or composition of set-ups, except for a few which have to do with the accommodation of editing demands. A shot of a flaming spark, six frames long and lasting a quarter of a second, can be a set-up; so can a ten minute, reel-long "sequence shot" which encompasses what would amount to a full scene in a theatrical production. (The comparison is apt since, with few exceptions, such shots are dialogue scenes in which the emphasis is on talk rather than on imagery.)

Although there are no rules for set-ups, there is at least one axiom: A set-up is valid as long as it is the best angle for showing that which the filmmaker wants the viewer to see. It has no other reason for being, aesthetic or otherwise. The instant another set-up is judged to be superior for the purposes of the scene, it becomes the cut of choice.*

It is obvious, then, that as a scene is being staged and rehearsed, the director must analyze it for its critical points of interest and transition, then tailor his set-ups to show them to optimum advantage. Every scene, every frame of film, presents a message, intended or not, and a moment of carelessness can uncover one which may be undesirable.

A chess master plans a number of moves ahead in his attempt to reach a favorable position; the filmmaker must also see far enough ahead to know, with some degree of certainty, where one set-up will be supplanted by another so that the actual *cuts*, which deliver a message of their own, can be made with a minimum of distraction or shock to the viewer. To facilitate this process many film theorists and a few filmmakers, Eisenstein among them, have recommended a structural approach to motion pictures. This methodology may be valuable in the analysis of a completed work, but it can be highly inhibitory if used when creating a sequence. A mechanical approach will usually achieve a mechanical result. In practice, each scene has its own dramatic imperatives, and when these are clearly recognized and properly developed it is surprising how often "laws of structure" can be deduced from the finished film.

Eisenstein said, "The strength of the montage resides in this, that

---

*When filming, the set-up, especially if it is a master or a long shot, is not cut off at this point, since it may reassert its primacy later on in the take. For various reasons shots usually begin before their usable starting points and continue after their probable cut-aways. For example, in the excerpt from *Casablanca* (Chapter 7, pages 63-64) there are eight cuts, but these are derived from just five set-ups. The three close-ups of Lisa, numbers 1, 5, and 8 are all segments of a single shot. Cuts 4 and 7 are also derived from one angle. These two set-ups probably run the full length of the scene, but only the appropriate sections were used as cuts. In many instances, one set-up may serve to supply a much greater number of cuts than in the example given. "Cutting in camera," that is, shooting only the section of a scene deemed usable in the final cut, is an extremely undesirable practice. (See Edward Dmytryk, *On Filmmaking* (Stoneham, MA: Focal Press, 1986), p. 425.)

it includes in the creative process the emotions and mind of the spectator [who] is compelled to proceed along the selfsame creative road that the author traveled in creating the image."

It follows that the message must be carefully particularized and isolated from contradictory, deceptive, or simply indistinct impressions. Ambiguous sign posts serve only to confuse the traveller, and a befuddled viewer is an inattentive viewer. If the filmmaker does not create the image that suits his exact purpose, a less effective, even a contradictory message may reach the screen.

The single most important aspect of the set-up is its *point of view*. Most shots—probably all shots—are the *viewer's* point of view (see the analysis of the *Casablanca* sequence in Chapters 7 and 8). But it is *not* the point of view of a viewer sitting in a particular seat in a particular movie house. One of the screen's most positive attributes is that there is no positional actor–viewer relationship. This fact liberates both the viewer and the actor. The actor does not play to the viewer; he relates and reacts only to his coinhabitants of the scene. On his part, the viewer is able to observe the characters and their milieu selectively from the most advantageous viewpoints the director can supply. He accompanies the scene's character as he crawls under the bed in search of an intruder, or under the desk to look for concealed microphones; he floats in the air to spy down a chimney, a well, or into a woman's pocketbook; he marvels at the symmetry of a mushroom cloud from many miles away, or stands nose-to-nose to look into a character's eyes while that character is unaware of the viewer's presence or the invasion of his innermost thoughts. The viewers watch a character's eyes as they focus on something offscreen, then instantly slip inside them to see the object of the character's attention. But viewers can do very few of these things in a "sequence shot" which, even with the advantage of camera movement, pins the viewers down in their seats by offering them only a limited point of view, while giving them too much to look at.

"The world is so full of a number of things, I'm sure we should all be as happy as kings."* In a poem or a sightseeing trip, possibly; in a film, not exactly—certainly not if the "things" assault the viewer's senses all at once. For, as Proust wrote, "In the most trivial spectacles of our daily life our eye . . . neglects . . . every image

---

*Robert Louis Stevenson, *A Child's Garden of Verses*.

The viewer assumes the perspective of the camera through all its movements, guided by the selection of the director. Gregory Peck searches Diane Baker's pocketbook in a scene from **Mirage**, a Universal film.

that does not assist the action of the play and retains only those that make its purpose intelligible."* How true, and how applicable, for the filmmaker must never forget his accountability to "essential" reality.

Fortunately, the motion picture is, above all, capable of isolating, and the filmmaker has not only the privilege, but the obligation to particularize his "view," to *exclude* non-essentials and those "things of the world" which are of no importance to his "vision." However, the camera and the microphone are not, by themselves, selective; each avidly records everything within reach. And it is only through arbitrary selection and the blessing of editing that the viewer can be made to see those details which, in his real world, would be winnowed out for him by his eyes and mind.

From a scenic environment replete with detail a director can isolate those objects or emotion-inducing images and sounds which best serve his purposes. From, let us say, a setting of a city street in the rain, melancholy can be evoked with shots of a thinly-clad, shivering, street person huddling in a desolate doorway, a feebly-flickering neon sign, a bedraggled bird hunkered under inadequate eaves; on the other hand, an emotion of quiet pleasure can be created by zeroing in on a child blithely splashing its way down a flowing gutter, a window box whose thirsty plants drink in the rain, and the warmly lit windows of a candy store. Selective editing of readily assimilatable set-ups can eloquently reveal a character's thoughts or at-the-moment state of being, because "the spectator is *compelled* to proceed along the selfsame creative road that the author travelled," and share the emotions of the author's choice. This in no way prevents viewers from exercising their freedom of interpretation, but it does limit their choice of *what* they are to interpret, and directs their emotional reactions along arbitrary paths.

Kracauer, in his *Theory of Film*, writes, "the spectator cannot hope to apprehend . . . the being of any object that draws him onto its orbit unless he meanders, dreamingly, through the maze of its multiple meanings and psychological correspondences." Perhaps early films, in which the existence of a remarkable "messenger" was itself the message, could intrigue the dawdler with long, busy shots, and send him into contemplation of his own related dream world, but such casual reception of a film's message is no longer possible, at

*Marcel Proust, *The Guermantes Way.*

least not in narrative films. It took little time for film pioneers to realize that viewers who withdrew into a dream world also withdrew from the film, and to recognize the need to direct the viewers' attention, to cue their train of thought, to keep them locked into a desired mood or emotion, to *lead* them through the "maze of multiple meanings." This recognition resulted, first, in the creation of "straight-on" close shots, then, in short order, in selective set-ups of many different kinds.

Every set-up, general or specific, must relate to the film's characters or the story. Seen for its own sake, an artistic shot may do more harm to a film than can be justified by its aesthetic appeal. Under certain circumstances, a beautiful composition in a close shot may direct the viewer's attention to an attractive *objet d'art* in the shot's background when what is wanted is an instant awareness of a delicate reaction in the actor's eyes. And a vista, whether in New Mexico or in India, should promote the viewer's understanding of the on-screen people who see it, whether they live in it or are just passing through. A long shot of a city or one of its streets serves no purpose unless it relates to the city's dwellers and any action that takes place in it.

An establishing shot, or almost any shot used merely to show the environment, has little meaning if the viewer cannot relate it to those who experience that environment. Relevance will be more immediately recognized and its presence in the sequence will be more meaningful if the viewer makes contact with the character first, then with the setting (see Chapter 3, pages 23-24). In other words, if it is clear *who* is involved in the full shot the viewer will more easily segregate and absorb those details which help to further the scene and develop the characters.

Some theorists and a few filmmakers prefer the "sequence shot," though the filmmakers can rarely afford to be purists. Without exception they find it necessary to take advantage of "intra-sequence cutting."* But, ideally speaking, one of the chief arguments in support of the sequence shot is its inherent "temporal realism." What a reactionary point of view! What the narrative filmmaker usually has in mind is the *avoidance* of temporal realism, as well as its corollary, "spatial realism" with its circumscribed frame of refer-

---

*A "sequence shot" is a long take which ideally develops a scene without benefit of cutting. "Intra-sequence" cuts are those made within a sequence shot.

ence. "Art," says Allan Bloom, "is not imitation of nature but liberation from nature."* The freedom to manipulate space and time is one of the great advantages of the film medium.

Experienced filmmakers learned more than a half century ago that a realistic temporal structure in most scenes is, with occasional exceptions, lifeless and boring. (Although it is a facet of staging rather than set-ups, it is relevant to mention here that even the pacing of speech and reaction to dialogue must often be speeded up beyond reality.) There may be an occasional viewer who prefers to decipher the *mise-en-scène*, to analyze the composition, to find his own center of attention, and to make his own dramatic selections, but such viewers are few indeed.

Narrative films do not, as a rule, tolerate this technique for a number of reasons. Viewers vary greatly in quickness of response and in their ability to observe, even to recognize, detail. If a director provides sufficient scene time to indulge the more leisurely and more thorough viewers, the minds which grasp detail and ensemble quickly or carelessly will probably find themselves bored and annoyed.

A series of shorter shots, each offering less, but more readily digestible material, equalizes the time required for absorbing and understanding the scene, while it sustains viewer interest at peak level as each new set-up discloses previously unseen frames and compositions. Collectively, it can deliver all, or more, of the details that a sequence shot might show, do it with superior arrangement, through editing, and do it in less time. A long, involved sequence shot inspires wonder and awe at the film crew's logistical expertise rather than its dramatic effect. Huxley's comment on theory has never been more apropos.

The truth is that the sequence shot directors—Welles and Renoir, among others—frequently resorted to intra-sequence cutting, normal cutting, and even Eisensteinian montage when the occasion demanded. And why not? The best filmmakers will always take advantage of those technical variations which will produce the greatest viewer involvement. In the classic *Sunrise*, Murnau's early scenes are laid in a rural setting, and are lit and shot to produce an ominous mood of potential tragedy. The concluding scenes, in the same setting, are similarly lit to realize a frantic mood of anger and despair.

*From Allan Bloom, *The Closing of the American Mind* (New York: Simon & Schuster, 1987).

But separating these two dark sections is a long sequence of unsurpassed gaiety which takes place in an urban setting. The lighting, the rhythm and pacing, the set-ups, and the cutting are all decidedly different from the manipulation of the same elements in the other two sections.

A good filmmaker will intuitively favor those set-ups which have molded his style, but he will never on that account alone neglect any set-up when an exceptional situation demands it. It is interesting to note, for instance, that a deliberate, relatively static dialogue scene will often require a montage treatment which is quite like that most effectively used in a sequence of physical action, simply to give it "life."

For obvious reasons action scenes customarily feature a succession of short shots, discrete bits of movement which can be combined to deliver a scene as a whole. For much the same reasons, so do involved conversations. By substituting the concept of mental movement for physical movement, one may find it as useful to explore in separate set-ups every change in thought, every new reaction of characters engaged in a verbal barrage, as it is to capture every change of position, every movement of attack or defense in a physical confrontation, though the paces of the two sequences might differ.

Using a variety of set-ups to create a variety of moods seems aesthetically logical, but once a sequence style has been established sudden stylistic change within the sequence will succeed only in wrenching the viewer out of his involvement in the scene. A striking example of such a result is seen in Elia Kazan's superior film, *East of Eden*. Raymond Massey and James Dean are engaged in a bitter father–son argument. Suddenly, for no apparent or understandable reason, the viewer is presented with a matching series of sideways-slanted close-ups. He probably assumes these skewed compositions are symbolic—but of what? It is difficult to relate them to the drama of the scene, to puzzle out the meaning of the arbitrary and apparently unmotivated tilts while simultaneously following the dialogue and action. To paraphrase Billy Wilder, "If you must use symbols, make them obvious." In this instance it is only the artful experiment gone wrong that is obvious, and meanwhile a good deal of viewer attention has been squandered in search of hidden meanings.

The choice of certain set-ups at either end of the spectrum—a brilliant, long-ranging move in ballet, for instance, or a close-up of a butterfly—is more or less obvious. Their effectiveness would depend on the proper choices of lens, lighting, and composition. But the choice of full or medium shots as opposed to close-ups, when either can be used, should be influenced by the requirements of the viewer. The situation may call for eye reaction only (close-up) or for the use of body language (medium or full shot). Eyes alone can reflect fear or express happiness, it is true, but only the body can "shrink back in terror," or "tremble with delight."

Almost without exception the best screen actors were and are masters of subtle body language. While analytically watching a Cary Grant performance (if that is possible) a viewer will soon appreciate a master at work. His body movements frequently say more than does his face; using too many close-ups in a Grant scene is like shooting a Baryshnikov ballet in chokers. And few old-timers are likely to forget John Wayne's walk or the seductive slink of Marilyn Monroe. The current overuse of extreme close-ups in almost all situations, subtle or not, has to a great degree deprived modern filmmakers of skillfully executed body talk, one of the most cinematic elements at their disposal.

The amount of information included in the scope of a set-up is a determining factor of its appropriate duration in time. With occasional exceptions, the impact of a close-up is immediate and direct; it needs little time to deliver its message. But as a shot increases in scope, so does the number of observable details which attract the viewer's attention, if only subliminally. That means, of course, that the pace of the fuller shots must be slowed down to allow the viewer more time to assimilate the additional details and to accept the intent of the scene. For example, in *Crossfire,* a lonely soldier who has inadvertently bruised the feelings of a cafe B girl (Gloria Grahame), asks her to dance with him, extracurricularly, in an unused garden section of the cafe. Somewhat reluctantly, she consents, and the viewer watches them dance to a Jellyroll Morton blues tune. There is no dialogue to diffuse the scene. As Ginny's dignity is slowly restored she relaxes her rather rigid stance. Soon she presses closer to her partner, and finally rests her cheek almost affectionately against his. By the time the music and the dance are done the viewer is not surprised to hear Ginny invite the soldier to share a spaghetti dinner at her apartment.

The changes in Ginny's mood and the accompanying attitudes are shown first in a medium two shot, then in a full shot. No close-

Medium and full shots are powerful, often overlooked cinematic tools. Gloria Grahame and George Cooper in Crossfire, *an R.K.O. Pictures film.*

up is necessary; body language says it all better. In truth, her eyes could not begin to express the slowly developing change in her emotions. But though her spontaneous physical moves are easily understood, they cannot be taken for granted, and a good deal of time is allowed to ensure the viewer's acceptance, emotional appreciation, and empathy.

A single dolly shot can follow a scene's characters as they move through a scene; the *moving master shot* is described in Chapter 6. But another kind of moving shot, one which isolates the character in his perambulations in a limited space, can be especially valuable. This is a talking shot, a monologue, in which the character, while ostensibly speaking to another character, really talks to the viewer.

Again, in *Crossfire*, a suspense film which deals with anti-Semitism, a police captain (Robert Young) tries to talk a frightened young soldier into helping him trap a murderer. When straight persuasion fails, the captain attempts to drive home the evils of racism by relating an incident of anti-Irish prejudice during the "know-nothing" era in American history. Starting in a medium two shot, the captain steps away from the soldier, and the camera moves into a loose single as he roams about the unoccupied area of his office, making his pitch. The set-up lasts for six or seven minutes, and varies in frame size from full shot through medium shot to, when necessary to drive home a vital point, a large close-up.

There is no cut-away, no inclusion of any one of the several characters who share the sequence. The director's intent is to avoid any displacement of the viewer's attention that might be caused by the sight of listening characters, and to induce him to forget for the moment that the speech is being made to influence the soldier by keeping that character off the screen. In fact, the speech is being made to influence the viewer, and for the greater part of the scene the message is being received by the the viewer in the soldier's stead until, as the scene nears its close, the captain sits down on the edge of his desk while the camera moves in to include the soldier in an over-shoulder shot. The viewer is now subconsciously and gently returned to the role of participant in the scene rather than the recipient of the message. And if he has been swayed by the captain's words, as most viewers are, he finds the young soldier's agreement to cooperate completely believable.

Two of the more important things that differentiate film from the other narrative media are the use of metaphor, though that is almost a forgotten art, and the set-up. It must be obvious that the set-up is a most powerful instrumentality for isolating attitude and action, for arbitrarily focusing the viewer's attention, and for heightening emotion beyond the pose or the dialogue. Because of its broad range of expression and impression and the nearly infinite opportunities for emphasis it offers the filmmaker who is knowledgeable and gifted enough to really use it, it remains unique in the world of dramatic art.

5

Invisibility

Art and ego have always been collaborators. Which does not mean that where there is ego there is always art—far from it. But, as Alan Jay Lerner once said, "Modesty is for those who deserve it." It has never overwhelmed a skeptical world, and without ego a work of art would never see the light of day. Happily, in great art the ego rarely shows; the artist's character, frequently; the ego, hardly ever. It may be that great art obscures the most flagrant of egos, or it may be that a real artist is never truly satisfied with his work. It isn't so much, "I should have done better"; it's, "What I really had in mind seemed more profound," or more beautiful, or funnier, or more frightening—whatever. Somehow, great dreams are always diminished in the realization, if only for the dreamer.

The "glamorous" professions have always attracted egocentrics in large numbers and, with the possible exception of politics, the film business has always suffered the greatest share. So it seems remarkable that the unwritten rule in filmmaking is, "Keep it invisible." The final implementation and best example of this precept is found in the cutting room, where the editor works hard and long to arrive at a "final cut" that plays as a seamless film in which cuts are not visible, staging is not obvious, camera manipulation is not noticed, and acting is not "performed." But, since staging, acting, photography, and cutting constitute the greater part of filmmaking, it follows that the most successful "tricks" of the filmmaker's craft are tricks of concealment. Just as one of a magician's indispensable skills is his ability to distract the viewer's attention from the "how" of his performance, so one of the director's greatest skills should be the ability to plan and execute staging and set-ups to the end that, however different in size, pace, and camera point of view, the scenes will ultimately meld into a seamless film in which the individual parts are blended into what appears to be an indivisible whole.

The preceding paragraph makes two assumptions: (1) seamless films are common, and (2) all filmmakers accept the principle of invisibility. Neither assumption is true. Perfect invisibility is impossible of attainment even by those who believe in the principle, and there are directors who seek instant appreciation of their technical virtuosity, which means it has to be instantly obvious and therefore trite. A few film editors have also been known to court attention with cuts that announced themselves—intentionally.

To make the point specific, here are two examples of filmmakers' tricks that break the rule. The first is an attention-getting line—a writer's trick not uncommon in films. The second is an attention-getting set-up which succeeds completely in impressing the viewer—but to what end? Both are well remembered and much admired. Both serve to draw attention to the filmmaking process. Both are excerpted from one of Hollywood's most respected films, *Gone with the Wind*.

1. As Rhett Butler walks out on Scarlett for the last time, he stops at the door, turns to face his tearful wife, and utters that famous line, "Frankly, my dear, I don't give a damn!" Today, the line would raise no eyebrow, but in 1939 the word *damn* was a cinematic taboo. However, its presence in the film was successfully fought through the censor board to the accompaniment of world-wide publicity. As a result, the audience reacted to the scene largely in that context, and the flow of an absorbing sequence was interrupted as viewers recalled the situation, considered the wickedness or innocence of the word itself, and commented on it to their seatmates. At that moment, the realization that this was, after all, only a film, was paramount in the viewer's mind.

2. Then there is the battle for Atlanta, and we follow Scarlett out of a shockingly overcrowded hospital into a huge railroad yard where the overflow of shattered humanity has been deposited on the bare ground—a wretched and pitiful sight. As she makes her careful way through the moaning and crying wounded, the camera, mounted on

A spectacular shot is not always the most desirable. A remarkable but questionably effective scene from Gone with the Wind, *a David O. Selznick Production.*

a special crane, moves slowly up and back, gradually disclosing a view of thousands of casualties. And long before the camera comes to a stop the viewers, without exception, are thinking, "My God! What a spectacular shot! How did they do that?!" There is even applause—great for the egos of all who created it. But the now forgotten message is, "Oh, God! What a tragic spectacle! What a dreadful waste of humanity!" Which of these reactions is ultimately better for furthering the story and its substance?

The paradox is that such a shot is considered one of the glories of film. Well, it is and it isn't. It all depends on how and when it's done. In the first example, Rhett's parting shot is the climactic moment of the film, the moment the viewer has been awaiting, the moment when Scarlett, who has been treating Rhett quite shabbily throughout the entire tale, finally gets her comeuppance. That's all it's about, and it is enough. It is not about Selznick's tug-of-war with the censor, and though it is true that the intrusion of this extraneous "subtext" leads to audience excitement, that excitement has little to do with the drama on the screen. Indeed, it serves to diminish it by intruding on the viewer's concentration and diverting his attention.

In the second example, the spectacle of the wounded could be shown, Eisenstein-fashion, in a well-designed series of shots—shots as significant as any part of the pull-back shot, and more effective from the *story* point of view. Although pull-backs and move-ins are frequently used to dramatic advantage, every effort should be made to camouflage camera movement. But in the given example, as the shot continues, the *act* of pulling back becomes more impressive than the subject of the shot itself. The *shot* becomes the message; it catches the viewer's attention and destroys the mood and the flow of the narrative. For a considerable length of time the viewer wonders, admires, and marvels at the "pulling of the strings," while the substance of the scene takes second place.

Contrary to normal expectations, it is the continuous moving shot which often alerts the viewer to the presence of technical contrivance. And though it may seem logically inconsistent, arbitrary cutting from one stationary set-up to another in a consecutive series of cuts—cuts which may vary in size, subject matter, and setting— frequently produces smoother results than would an uncut master shot. In an edited sequence, each cut furthers the narrative and foreshadows the next appropriate thought or action as timed (and this is important) for the screen. The next cut may be what the viewer expects or wants to see in the unfolding plot; it may be a surprising turn, a contradiction, or an unanticipated development that the filmmaker is ready to share with the viewer; it may be a balancing of movement or attention that is a part of a sequence involving parallel or interweaving action, as in a chase or a "ride to the rescue." But in all such instances, if each cut develops story or character at a desirable pace and the film editor takes advantage of the smooth cutting techniques at his command, a dramatic involvement can be achieved which blinds the viewer to the changes of cuts. And, of great importance, *only* cutting makes it possible to manipulate the viewer's attention, to direct it exclusively to the precise parts or people that the filmmaker wants him to see at any particular instant. The viewer's attention is given no time to stray.

But first, let us examine a simple camera move—a basic film expedient long used for an opening shot in a sequence. A dissolve or fade-in discloses a brightly burning campfire, we hear offscreen music, then the camera pans from the fire to a lone cowboy sitting nearby, noodling on his harmonica. Although the music helps to unify the scene, the movement of the camera is unmotivated; the viewer is at least momentarily aware he is watching a "setting." This cliché technique simply introduces the scene and sets a mood.

However, if we can arrange to have a flaming ember—perhaps a scrap of burning paper—blown out of the fire by a sudden gust of wind (and we can) and a skillful camera operator keeps the eye-catching spark in the picture as it flies toward the cowboy, no self-consciousness is present. The flare-up and flight of the flaming spark gives the fire a reason for being—it is a player in the scene, not a prop—and the moving background is unnoticed by the viewer whose attention is fixed on the screen-centered spark which, in reality, does not move at all in relation to the screen. The effect is as if the image of the cowboy dissolved subtly out of the scene of the flying spark. And the viewer's involvement may be increased as he suddenly finds himself concerned with the possible results of the meeting of the flying ember and the cowboy's coat.

A set-up in which the camera moves of its own volition is not only self-conscious and, as such, a sign of technical inexpertise, it distorts the concept of narrative filmmaking as well. For example, picture a group of four characters in conversation. Character A makes a challenging statement; the camera pans to character B for her reaction, then to character C, and eventually to character D who, by this time, is hopping from one foot to the other. Such a shot cannot claim a single advantage over a series of reaction cuts, but it does exhibit a few faults.

1. B's reaction must be delayed until her face occupies the screen and would, unless B has been established as a slow thinker, appear to be ill-timed. The timing discrepancy would grow progressively greater for C and D.

2. Since there is no acceptable way to synchronize the flick of an onlooker's eyes from A to B to C to D with the much slower pan of the camera, the mechanics of its movement would force itself on the viewer's attention.

3. The time involved in the pan (or pans) is extra baggage, not one second of which is useful to the film.

4. Fixing the timing in the pan shot eliminates the possibility of editorial refinement, which is akin to skipping the use of sandpaper in the finishing stages of a fine piece of cabinet work.

The question has been begged here by the use of the worst possible example of camera movement, an example, however, still too often seen on the screen. But camera movement, which was seen as a great boon when first developed by men such as Murnau and Lang, still presents numerous problems for the meticulous filmmaker, not only problems of timing, but problems involved in the aesthetic development of the art of filmmaking as well.

For movement, too, must be "invisible." If viewers are conscious of camera moves they are obviously watching technique. Yet the mark of a true artist, whether tailor, architect, or filmmaker, is that "the seams don't show." If the difficulty involved in the unsurpassed digital dexterity of Art Tatum or the amazing physical virtuosity of Baryshnikov were obviously displayed in their performances neither name would live in the history of art. And if a film director allows his "seam" to show he is either careless, inexpert, or seeking admiration for his "brushwork" rather than for the worth of his statement as a whole.

Good films accentuate the cinematic by using images that convey meaning, as opposed to relying on dialogue. A scene from The Young Lions, *a 20th Century-Fox film.*

6

Moving and Molding

One of the unique advantages of film is its malleability—its capacity to respond to corrective measures, to create perfect fragments that can be welded into a perfect whole. How ideal—and how impossible. Because, of course, there are a number of limiting factors; the quality of the story, the personalities and talents of the actors, the temperament, skills, and intuition of the director, the competence of the crew, the ability of the film editor, whoever it might be, the time and money available to accomplish the work, and a host of lesser hurdles. One cannot arrive at an equation, construct a theory or even a concept of filmmaking without taking all these variables into account. The difficulty is that there are so few constants to work with.

The cinematic idealist sets out to create a theoretically perfect set-up, and runs into literature. Almost without exception, the good screenwriters are exactly that—writers, or playwrights. Today, one rarely reads a script that would tempt Murnau or Lang or Von Stroheim or even Lubitsch, who was essentially a man of words. Many filmmakers who would like to accentuate the pictures in "talking pictures" rather than the talking are disconcerted by the literary character of their scripts. Eventually, they settle for motion, as in "motion pictures," in the mistaken belief that movement will do the job.

But, although the two sometimes coincide, physical movement should never be confused with imagery. Movement alone means little; it will not, of itself, rescue a talky scene, and it may muddy a good one. A good dialogue scene is an interplay of ideas, of verbal conflict, of persuasion and reaction (never, if possible, of direct exposition). It has a movement of its own; the movement stimulated in the viewer's mind. Physical movement is not necessarily cinematic per se, and an unsatisfactory dialogue scene will rarely be improved by its imposition. The only sure cure for a bad scene is a better one—if possible, one that says largely in images what the writer has ineptly said in words, and the effort to create such a scene should always be made. But that's asking a lot, and most directors will fall back on a compromise; they will "touch up" the dialogue a bit and devise a staging which, they hope, will add vitality to the scene through choreography.

To be sure, there are situations in which extensive movement accompanying speech is advantageous. For example, in *The Young Lions*, Hope's father, a scion of old New England stock, meets her financé, Noah, and learns that he is from New York, poor, and a Jew. The father is not pleased. He is a man of untested principle, but decent, and he takes Noah for a walk around the small town square. Without obvious comment, he points out the old office buildings and stores which house fifth and sixth generation Vermonters, the church cemetery where several generations of Hope's family lie buried, and the old school where more generations of WASPs have learned their ABCs. Here the walk works for the viewer; subtly, engagingly, cinematically. The father's indirect but unmistakable effort to convince Noah that he and Hope are a social and cultural mismatch is much more effective than a direct attempt at verbal persuasion, and a good example of a proper wedding of words and images. But a long dolly shot of two people elbowing their way along Fifth Avenue, shouting lines which could be better said and more calmly considered in a quiet room, is a waste of time and money. "Taking it outside" does not, in itself, make a scene more cinematic, and adds nothing to the film if it doesn't belong there.

The tendency to play it close, play it long, but move it, has developed slowly over the years into the modern master shot, a useful set-up which frequently leads to sloppy filmmaking. During the thirties, directors working around the cinematic restraints of the new "talking pictures" developed a staging style that was in some important ways the antithesis of the best silent techniques of the twenties. Filmmakers like Capra, Wyler, and Stevens took little

For years, long, unbroken stretches of dialogue were a significant part of successful motion pictures. Clark Gable and Claudette Colbert in a scene from It Happened One Night, *a Frank Capra film for Columbia Pictures.*

advantage of the camera's versatility and mobility.* Capra, especially, seemed to care little for technical virtuosity, but his stories
were so engrossing and believable, his dialogue so entertaining, his
humor so rich, his characters so absorbing, and his pace so finely
tuned, that the viewer had neither the time nor the desire to analyze
the film's technical shortcomings; he could only enjoy.

With the exception of John Ford, the top American filmmakers
of the period could, and often did, shoot unbroken scenes in which
two immobilized actors, usually in profile, talked to and at each
other for many minutes on end. Technically archaic, but who cared?
The quality of their characters and the words they uttered usually
carried the day. Then came Orson Welles. He had not yet made a
film, but he loved the old, silent masters, and he soon reminded the
world's filmmakers of their cinematic heritage. He showed them
they could accentuate the good things they were doing by using the
camera to create rather than simply to record.

Wide angle lenses, skillful lighting, and creative camera positioning now informed the viewer that rooms had ceilings and floors
and foreground pieces and depth and dark corners. Moreover, the
camera could move with the actors and diminish the risks of the
film's passage through the cutting room.

Fear of such risks occasionally influences the choice of shooting
techniques. Most directors have never had any hands-on editing
experience, and some of them are uncomfortable with the cutting
process—with reason. The cutting room is where much of the film
is manipulated, and mistakes can be made there as well as on the
set. Every good filmmaker knows what he wants in his final cut,
but knowing what he wants and knowing how to get it are two quite
different things.

This editorial insecurity frequently forces directors to shun one
of film's most fertile attributes: shooting in bits and pieces. They
believe that the longer take offers the actors greater opportunity for
"getting into" their characters, and that they, themselves, can better
judge the scene as a whole. The current prevalence of shorter sequences and more dialogue plus the ready availability of excellent
camera moving equipment makes it relatively easy for the director

---

*In 1940, I was filming a Karloff suspense film, my first B for Columbia. I requested
the "big crane" for my opening shot. The studio's reply was short and to the point:
"*Capra* never used a crane!"

to sublimate his fears and insecurities by staging and shooting the scene in a single set-up. When all the elements are right this can be a very useful and generally thrifty shot. It can even, in rare instances, be brilliant, as the opening set-up in Welles's *A Touch of Evil* demonstrates.

Unlike its progenitor, such a master shot need not begin as a long shot; in fact, it rarely does. And once under way, it can wander around the entire set, or a large part of town, limited only by the requirements of proper lighting and the capacity of the film magazine. It can pull back into a full shot or creep up into a choker; actors can walk up into close-ups or fade away into the sunset. Let us stage a hypothetical scene—only the movement, no dialogue.

The shot begins with a *close-up* of a ringing telephone which rests on a side table. The viewer does not yet know that he is in a living room, but he soon will. On the third ring a woman's hand enters the shot and picks up the instrument. The camera *pulls back* and *pans* with the movement of the arm, following the phone into a *close-up* of the woman. During the short, one-sided conversation, she casually takes a few steps away from the side table (for reasons which will shortly be clear) and the camera *moves with her*, maintaining the *close-up*. Her telephone chat is soon interrupted by the sound of a door bell. She glances toward the offscreen door, excuses herself, and the *stationary* camera now *pans* her back to the side table, where she puts the phone down without hanging up. Since we dollied with her during the phone conversation, her walk back to the table, away from the camera, has moved her into a *full-length shot*. Now the camera *pans* with her as she crosses the living room into the entrance hall and toward the front door. Perhaps it *follows* her, slowly, as we prepare for the next move, but now it frames her in a *long shot* as she reaches the door, opens it, greets her visitor, then leads him back into the living room. The camera simply *pans* with them until she stops at the couch, invites the caller to sit down, then drops down beside him. While this action is taking place, the camera *moves toward* the two, dropping down a little, and ends up in a *close two shot* or, more usefully, a close *over-shoulder shot* of the dominant character in the scene, which continues to play until we cut away to another scene or until the camera slowly, suspensefully, swings off them and onto the telephone which lies on the side table. Do we hear the buzz of a broken connection, or the silence of someone still listening?

Long dialogue scenes with little accompanying action are the rule today. For such scenes the master shot can be a useful set-up, and a time-saver—but not always; even a fairly simple moving shot can present problems. The more complex the set-up the more hands are required to make it work, and the more hands involved the greater the chances of error.

A satirical scene in *Day for Night* brilliantly illustrates a few of the difficulties inherent in this technique. In his film within a film, Truffaut shoots a master shot in which an aging, insecure, and over-compensating actress, who is also a bit under the weather, goes through her paces, losing her lines, muffing her moves, then making a confused exit into a china closet instead of through the proper door. After several botched-up takes she collapses in an emotional outburst. Though somewhat of an "in" routine, the scene is sad but amusing. In the real world, however, it is only pathetic. Occasionally, actors do forget their lines, even when sober—some because of advancing age, some because of other natural causes—and fear of failure dogs their steps. Under such circumstances a compassionate director will sacrifice his self-indulgent set-up in favor of several shorter, more manageable and, for the actor, less taxing shots. If he really understands filmmaking he will be more than repaid for his consideration. He will get a better scene, *because an important source of the value and the delight inherent in film is its facility for examination, for impressing with the big and intriguing with the small, for capturing the most subtle response to an action or a word.* The all-in-one master shot is often unequal to the task. It is no more than a scene from a play, only closer. It is incapable of showing different facets of a subject, be it a thing or a person, from several points of view that, when properly manipulated, coalesce into one.

However, the master shot *can* be modified to eliminate its faults; one need only take advantage of the art of film editing. The solution is to make the master shot the "spine" of a scene, not its entire body. In this role it is similar to the old master long shot, yet different, and better; it is amenable to the insertion of closer shots, yet never pulls away from the heart of the scene, as a long shot inevitably does.

Crossfire was a chancy film; its subject, anti-Semitism, made it a box-office risk. To reduce the possible financial loss it was made with little money and in very little time. The film was shot in twenty days and, in what turned out to be a very successful effort both critically and financially, a total of only 140 set-ups, for an average of 7 shots per day.

Perhaps a third of these were master shots, most of them running three minutes or more (one ran for ten). Spliced together they could have told the entire story, but a great deal would have been missing. The other two-thirds were largely close-ups and over-shoulder set-ups, reaction shots which brought out the nuances that the master shots had missed or glossed over.

The master shot can be useful for scenes which are completely dominated by dialogue and in which the reactions are relatively unimportant, but when imagery and reactions play the dominant role the value of the unprotected master shot is questionable—with a few important exceptions. First is the mood-building shot so common in mysteries, suspense, action films, and, occasionally, in high straight drama. In such set-ups the camera follows a single character as she or he moves through an empty warehouse, a dimly-lit alley, or any of the dozens of variations of these settings, where the smell of danger, real or imagined, hangs heavily. Speed of movement may vary from casual to frantic, the character can be aware of the danger, frightened, and reacting accordingly, or the lurking menace may be known only to the viewer, who will cringe in the actor's stead at every unfamiliar sound or wavering shadow.

A relatively short scene can easily be accommodated in one unprotected set-up, but an extended sequence, such as the one recording Simone Simon's suspenseful walk along an eerie, moonlit street in Jacques Tourneur's *The Cat People*, may require a series of master shots, which may be stretching the definition a bit.

The director who has no flair for cutting, and fears the editor, can resort to a technique which "invisibly" knits a series of short moving master shots into a complete sequence. First, the sequence must be broken down into several master set-ups, each concerned with an integrated portion of the whole. To effectuate a flow from one section to the next, each set-up may begin as a point of view (POV) related to the previous shot's final frames. For instance, as the first set-up (I) ends, the camera moves into a close-up of character A, whose attention is drawn off-screen. On his "look-off," a precisely timed cut to shot II shows the viewer that the object of A's interest is character B, who is just entering the room. This POV shot is the beginning of a second moving master shot, which may eventually be integrated with shot I, or introduce its own characters and material. In its turn, shot II flows into shot III, either through another "look-off" or by following a character from the B group as she moves out of shot II and into shot III. This series of short, moving master shots is continued until the overall sequence has been accommodated.

The advantage of such a series, containing two or more set-ups, is that each shot can concentrate on "telling it best." If smoothly cut even the "sequence shot" buff (see Chapter 4) will see a continuous shot, while the "normal" viewer will enjoy a scene shown to its best advantage.

However, the moving master shot may play its most subtle role when the actor in the scene apparently talks to an offscreen character while, in reality, he is delivering a specific message to the viewer. In such cases the presence of other actors on the screen would be completely unproductive. Here, the rationale is similar to that used in the better film musicals. Cuts of the on-screen audience may appear at the beginning of a staged number, but once the number is under way further shots of the on-screen viewers are detrimental; they would imply that the entertainment is being staged for their approval when, in fact, it is the real viewer who is now asked to judge its quality.

But where a musical number can consist of many cuts (even a solo turn usually incorporates a number of angles and occasional special effect shots), in a dramatic scene of this sort any straight cut, especially a cut-away to another actor, will disturb the line of communication between the speaker and the viewer.* In such a situation a skillfully executed master shot is the ideal set-up; a shot which moves and pans with the actor, but allows him or her to slip away into a full figure in casual moments and welcomes him or her back into close shots, even close-ups, when the scene needs punctuation.**

The moving master shot achieves movement of the mind through movement of ideas, movement of the film's characters, and the movement of the camera. However, a powerfully sensory movement which borders on the physical can sometimes be achieved by creative film editing even, perhaps especially, when the characters are quite motionless. The next chapter will demonstrate this particular technique.

*For a more detailed analysis of a specific shot see Chapter 4, page 43.
**This set-up is sometimes called a *follow shot*, but when it encompasses the meat of the scene, *master shot* seems a more fitting appellation.

7

Look at Him, Look at Her

Before the advent of film schools most of the terms used by theorists and academicians were quite unknown to the great majority of Hollywood filmmakers. Buster Keaton, one of comedy's greatest technicians, never learned the rules of grammar, let alone the principles of cinematic syntax. And even today expressions like *parallel action, separation, multi-angularity, Z-axis,* and more than a few of their kin are rarely heard on a set or in a cutting room.* But they do serve a purpose; they point to the fact that a great deal of what is considered to be the art of the cinema has little to do with substance or performance; the maxims of the art are discovered primarily through study of the film editing process.

In film's formative years the strange and special problems of filmmaking puzzled those whose knowledge and experience were derived from other disciplines. Eisenstein came to films from the theater, and he brought it with him. Even in the latter part of his career, when American and British actors like Tracy, Bogart, Colman, and Niven were "throwing away" their lines, Eisenstein's professional actors were theatrical to the point of hamminess. His scenes, especially his long shots, betray his background as a set designer; they are masterpieces of still composition. An enlargement

---

*At a seminar I shared with Jean Renoir, I was amused and a little surprised to hear him plead ignorance of the esoteric nomenclature the film students threw at him. I was also pleased that we were in the same boat.

of almost any single frame would receive a warm reception at a showing of photographic art. By film standards his movements are often disorganized and confused. And although he could create exceptional sequences like the "Odessa Steps," most of his scenes are static. As a young film editor at Paramount Studio I watched many of the rushes of his Mexican project. They consisted largely of innumerable shots of agave and colorful peons, composed in classic pictorial style. If there was any *filmic* action it escaped my attention.

If there is one rule that should hold for film it is that the techniques of filmmaking *must* be at the service of the material filmed rather than the other way around. It seems to me that many film scholars have based their theories on principles of classic art (including literature) rather than on an understanding of the different, *mobile,* art of the screen. It also seems to me that the art of film editing was developed *inevitably* to correct that point of view, to bring *cinematic vitality* to scenes rendered lifeless by their fidelity to the principles of the more static arts. One of the better examples of this contention appears in Eisenstein's drama of the modernization of a collective farm, *The General Line.*

In brief, a machine which separates whole milk into skim milk and cream is being introduced to a group of reactionary peasant farmers. In dramatizing his get-with-it message, Eisenstein completely ignored the fact that cream separators were common in the developed countries, that they were probably known and used in the more up-to-date areas of Russia, and that their ability to perform was beyond question. He and his writers cleverly contrived a rather long and suspenseful sequence—will the machine work, or is it a fraud? Can it really split milk into its main components? The viewer is kept guessing for several minutes while he watches the expressions on the peasants' faces change from curiosity, to skepticism, to derision, to sudden surprise, and finally to joyous acceptance. But the *changes* are all offscreen!

The cuts which are used to show *emotional movement* really don't; they are nearly all static. We never see a peasant *break into* a skeptical smile or *burst out* in derisive laughter. Instead of *onscreen* transitions we see the fixed results of *offscreen* changes in a series of portrait-quality close-ups which are intercut with close shots of the gleaming machine, whose chief moving parts, the spinning flywheel and the whirling milk-containing centrifuge, rotate so rapidly as to appear frozen. Even the commissar who has brought the machine to this backward collective, and who must know that it delivers, mirrors, in several close-ups, the peasants' doubts, near disappointment, and ultimate triumph.

The paradox here is that almost all the "reaction" close-ups show no reaction; they could have been frozen frames of still photographs. True reactions are rarely seen, only the end results, fixed expressions such as might be found on a placard illustrating "joy," "fear," "anger," for example. The transitions from curiosity to doubt, to derision, to joy, take place *offscreen* while the viewer's attention is on the machine, the young man who turns the crank, or the anxious commissar. When the viewer's attention is returned to the peasants the desired changes in emotional attitudes have already taken place. But, contrived or not, the sequence works. Creative editing, a skill at which Eisenstein had few equals, provides the movement which infuses the static scenes with a collective life.

This technique permeates Eisenstein's work, and is at the root of his system of montage. *Ivan the Terrible* is a gallery of portraits and tableaux. So, to a great extent, is *Alexander Nevsky*. The technique may have been developed to deal with the non-actors he so often used. After all, the art of *screen* acting is in the listening and in the resulting reaction, not in the pose or the pretense, and few amateurs, or long-time theater actors, for that matter, can handle it.

For a more modern use of close shot reactions let us examine a series of cuts made by film editor Owen Marks in Michael Curtiz's *Casablanca*.* At the piano in Rick's Cafe Americain, Sam (Dooley Wilson) starts to sing "As Time Goes By," obviously an old favorite. After four bars there is a cut to a large close-up of Lisa (Ingrid Bergman).

	Shot	Length in seconds	Total
1.	C.U. Lisa.	27	27
2.	Full shot, Rick (Humphrey Bogart).	7	34
	He enters the room—reacts— heads for Sam.		
3.	C.U. Sam.	2½	36½
	He continues singing, oblivious of Rick.		

* *Casablanca* can hardly be called modern, but nothing in the last forty or more years can be called a better example.

Humphrey Bogart and Ingrid Bergman in a scene from Casablanca, *a Warner Brothers–First National Picture.*

4. Dolly shot, Rick.	5	41½

He reaches Sam, who warns him
   with a look.
Rick looks off; sees:

5. C.U. Lisa.	3	44½

She is looking at Rick (offscreen).

6. C.U. Rick.	3	47½

A slight reaction. He continues
   looking at Lisa (offscreen).

7. Two shot—Rick and Sam.	2	49½

Sam beats a hasty retreat.

8. C.U. Lisa.	2	51½

Continues looking at Rick
   (offscreen).

The mood and the situation are interrupted by a full shot cutaway to others, who are then panned over to Lisa's table, and the sequence is off on another tack.

Before this series of eight cuts the viewer has learned only that Lisa knows Sam and his boss, Rick. There has been no information of a personal nature, no exposition or explanation; he only knows that Sam, for some undetermined reason, is wary of Lisa, and has lied about Rick's presence in the cafe. This brief section hinted at an awkward earlier relationship, nothing more. When Sam, at Lisa's urgent request, starts singing what is obviously a meaningful song, we see the first cut of the series, a large close-up of Lisa, angled out 45 degrees to the left of camera center. She does not react sharply, she does not "make a face." Her eyes glisten with a hint of tears, but only a hint. With virtually no change, the cut runs for 27 seconds—and accomplishes a great deal.

The length of the shot is extremely important. It gives the viewer time—time to start making sense out of a puzzling, nonexpository situation. He becomes aware that the song, which has not been previously heard in the film, probably falls into the "our song" category. He recognizes Lisa's pensive expression as something more than mere nostalgia. On the basis of that expression and that song he begins to sort out the hints and put them together—a superb example of encouraging the viewer to think, of drawing him into the story, of creating an active participant out of a passive spectator.

The second cut runs 7 seconds. The time has no special significance. 7 seconds is how long it takes for Rick to enter the room, become aware that Sam is singing a particular song, let his demeanor inform the viewer that it has a special meaning for him, and then to exit the shot, with determination, toward Sam, who is offscreen right.

Cut 3, 2½ seconds long, is a close shot of Sam, singing. It is used here as a filler to get Rick (offscreen) across the room for the next cut. It also serves to inform the viewer that Rick's attention is only on Sam; he is not yet aware of any unusual presence. A shot of Lisa at this point would have been more than premature; it would have destroyed the drama of the situation and weakened the emotional build-up.

Cut 4, a waist-figure dolly shot, 5 seconds long, brings Rick into the piano. With suppressed anger he starts to speak to Sam (the only bit of dialogue in the series) but stops in mid-sentence, halted by the singer's expression. Sam's eyes shift toward offscreen right. Rick's eyes follow Sam's look, then focus intensely on something or someone offscreen. Although there has been no set-up holding both principals in the same shot, nor will there be, the viewer understands that Rick has caught sight of Lisa.

Cut 5 is a *large close-up* of Lisa. It is 3 seconds long. It is not shot from Rick's point of view, though it serves as one; it is the *viewer's* point of view. Lisa is looking at Rick, offscreen left, and even though she has not been seen spotting Rick, the viewer will take for granted that she has been watching him since he entered the room. There is no movement of Lisa's face, only tears in her eyes; eyes which say everything the viewer needs or wants to know, and 3 seconds is all the time he needs to absorb it. Just when the viewer looks for Rick's reaction—he gets it.

Cut 6, a large, matching close-up of Rick, looking camera right, eyes fixed on Lisa (offscreen) is also 3 seconds long. It is *not* 3 seconds long to match the previous close-up, but because it takes that long for Rick to show surprise, then a subtle and quickly suppressed disapproval. When the reaction is complete, so is the shot.

Cut 7, a two shot of Rick and Sam, catches the singer already moving the piano and himself out of the picture. Only 2 seconds are needed to indicate Sam's exit, which clears the way for the dramatic scene the viewer fully expects will follow. Sam stares off at Lisa throughout the shot.

Cut 8 is a close-up of Lisa, actually a continuation of the shot used for cut 5. She continues to regard Rick for 2 seconds. But before the situation can be pursued, the scene and the mood are disrupted by a cut-away to a new set-up with other characters.

This series of cuts, aesthetically quite restrained and extremely economical, is as fine an example of a truly unique facet of film-making, the close-up reaction, as can be found in any film. It is also a perfect example of the art of getting information across with subtlety and indirection. And it addresses one aspect of film theory, the attempt to arrive at structural formulas. Here, there is no set formula, no "painting by the numbers," no "sets of three," no sets of any kind, actually, and no rules except the first rule of film editing: Each cut shall run long enough, and just long enough, to deliver its message. It is interesting that Lisa is seen only twice after Rick's entrance. It is enough.

Although more than half an hour of the film has elapsed before Rick and Lisa come face to face, and nothing in that 30 minutes has informed the viewer of a previous relationship, it takes less than one minute of largely close-up reaction to let him know that the two have shared a deep and memorable love at some (for now) unknown time in the past.

A sequence of this kind, which contains no obvious or broad reactions, yet says a great deal through what filmmakers call "looks," must take advantage of only the finest screen acting; acting that isn't "acting," where the eyes tell the emotional story though the face barely moves. A person off the street, however willing, or an actor conditioned to read the lines, however beautifully, will rarely fill the bill.

A shorter series of over-shoulder close-ups from *The Young Lions* illustrates the point in a different context. Hope (Hope Lange) is visiting her husband, Noah (Montgomery Clift) in an army prison where he is being detained for going AWOL. Seated, they face each other through a heavy glass partition, and talk through a speaker system. Because of this arbitrary positioning, the sequence was shot in a series of over-shoulder shots, each strongly favoring one character or the other. After a few cuts in which they speak briefly of Noah's predicament and the difficult decision he must make concerning the army, he pauses.

The over-shoulder shot is one of film's most useful angles. Montgomery Clift and Hope Lange in a scene from The Young Lions, a 20th
Century-Fox film.

1. Over-shoulder close shot, Noah.   7½ seconds.
   He studies Hope curiously, then speaks, "Hope,
       stand up. . . ."
2 O.S.C.S., Hope.   8 seconds.
   She looks at Noah uncertainly, hesitates.
   Again, he says, "Stand up!" She rises slowly,
       until her head is out of the shot.
3. O.S.C.S., Noah.   2½ seconds.
   His face lights up. "Honey. . . ."
   He starts up.
4. O.S. two shot—Noah and Hope.   5 seconds.
   As he continues to rise, "How long is it?"

In only 18 seconds, and no on-the-nose dialogue, information important enough to influence a vital decision is given through eye-contact in *separate* close shots. The viewer knows that Noah will voluntarily return to his unit and its brutal captain.

In cut 1, 7½ seconds long, Noah studies Hope's face and has a hunch. His "Stand up" is followed immediately by cut 2, an 8 second over-shoulder close shot of Hope. She hesitates, feeling he's had enough trouble. Noah repeats, "Stand up!" Her expression is non-committal as, facing full front, she rises slowly until her head is out of the shot, and the camera dwells briefly on her midsection. Since she wears a shapeless black dress and a light top coat, no stomach bulge is evident, but the fact that her head, usually the center of attention, is allowed to leave the frame while the camera remains on her torso, makes it possible to deliver a clear message.

Cut 3, an over-shoulder close shot of Noah, is a little more than 2½ seconds long; just long enough for him to break into a beaming grin and jump up to join her in a longer, standing shot. As he moves he starts his next line, "Honey . . ." which is completed in cut 4, a loose over-shoulder two shot, 5 seconds long. His line "How long is it?" and her answer "Five months" only confirm what the viewer has just learned. It is simply the icing on the cake. Hope's subtle expression and the unrevealing but informative shot of her stomach has blown her secret by the end of cut 3. Again, no structure or rhythmic effects. The rhythm of the scene is in the staging, the playing, and in the film editor's judgment of just how much information a viewer needs and gets from every cut, and in his superior *instinct* for the viewer's limit of forbearance.

These three scenes demonstrate the image's superior potential for dramatizing a scene, and the opportunity it offers the viewer to *think* along with the film's characters. A brilliant line of dialogue can evoke a viewer's admiration but it will not allow the time needed to analyze it, since more dialogue will nearly always intervene. But a silent shot, or one with purely supporting dialogue, will encourage the viewer to interpret, to think, as the scene unfolds, and thus more fully grasp its theme and its emotional thrust.

8

The Art of Separation

An interesting aspect of the evolution of filmmaking is the speed with which viewers have assimilated the changing conventions of the craft. Creative filmmakers once felt compelled to move carefully when dealing with innovative techniques lest they lead to confusion. Now they must use all their cunning just to keep abreast of the average viewer. Even obvious paradoxes of a medium sired by incongruity and born of contradiction are taken in stride. The most casual filmgoer accepts the fact that the "moving picture" is an illusion, a succession of tiny transparencies, no more than an inch square, which, at the instant of their projection on the screen must be absolutely still, and that the giant reflections on the flat screen are merely two-dimensional reproductions of people and things. However, overriding all this is the viewer's willingness to assume that these reflections are three-dimensional, and more real than most "live" representations.

Some of the more intriguing inconsistencies are those of visual dysfunction. One of the incongruities that help to individuate the medium is brought out in the example detailed in Chapter 7. The apparent inconsistency is this: In a scene enbracing two people it is *not* necessary to establish them in a two shot or a master in order to inform the viewer that they are, in fact, in each other's presence. Indeed, the cinematic *separation* of the two will always draw the viewer closer to them as individuals and as a pair. The contradiction may need a little explication.

In the *Casablanca* excerpt, the paths of Rick and Lisa cross for the first time in the film, an event the viewer has been anticipating. It would seem dramatically logical that here, of all places, the two should be shown in an inclusive set-up to establish their positions relative to each other. Yet, in this short sequence they are not seen together at all, for good and sufficient reasons.

First, the long close-up of Lisa which gives the viewer time to put things together also establishes the song, "As Time Goes By," as a motivating factor in the sequence. When Rick enters the room in a full shot, hesitates, and looks offscreen right, the viewer knows that there are two principals, Sam and Lisa, in that direction, and to remove any ambiguity a routine cut would have shown Sam as the object of Rick's look. But here, creative editing assumed that the viewer would accept the song, and therefore the singer, as the focus of Rick's attention. The assumption rendered the cut-away redundant, and the shot stays with Rick through his look-off, his angry reaction, and his exit from the set-up.

However, the "filler" close shot of Sam which follows and serves to give Rick time to cross the room (offscreen) eliminates any possibility of confusion. It also intensifies the viewer's emotional involvement by postponing, if only for a few seconds, the anticipated confrontation.

(It is apropos to note here that in a well-constructed, well-edited film, a cut introduced for one purpose will often serve other purposes as well, sometimes to the editor's pleasant surprise. The cut of Sam, inserted to bring Rick across the room in less than "real" time, also functions as a point of view shot and a suspense-building hiatus. One would be hard-pressed to say which purpose is actually the most important.)

When Rick does see Lisa, she is not shown in the traditional point of view shot which, following the rules of film perspective, would be an angle closer in size to the two shot of Rick and Sam; she is shown in a huge close-up, a head. Strictly speaking, this shot does not show the viewer what Rick sees, but what the viewer sees Rick *looking* at. It is the viewer's point of view, not Rick's, and its purpose is to extract *from the viewer* an emotional reaction quite apart from, and possibly contrary to, that which Rick feels; to evoke the viewer's response to the emotion in Lisa's eyes. If the viewer is not touched and, at this point, does not want more than anything else to see Rick and Lisa together, there is no point in continuing the film.

If asked why he cut a sequence in this fashion, a creative cutter would cite no rules; he would probably answer, "It just seemed the right thing to do." But in analytical circles, the *Casablanca* excerpt would be seen as a close relative of the technique known as "separation"; a cutting procedure most frequently employed in sequences of extended dialogue. The number of characters involved is usually two, but it can vary; a sequence in *The Carpetbaggers* moves to its climax in a series consisting of 14 consecutive close-ups of four separate characters, * but this would be considered somewhat uncommon. The functions of the procedure also vary. It is used in solving problems of pace, of rhythm, and of emphasis, but the only factors to be considered here are the engagement and the participation of the viewer.

The late Viennese architect, Adolph Loos, wrote, "The work of art is the private affair of the artist. The house is not." A slight variation makes an ideal motto for the filmmaker: "The creative quality of a film is the private affair of the filmmaker. The film is not." A film which pleases its audience may please its maker, but the film made *only* to please the maker and interest a few theorists is a total waste of time. It is essential to engage the viewers in this particular work of art, and the only way to judge the extent of their engagement is to measure their participation in the film's close-ups.

Let us set up an imaginary sequence in which two characters are holding a conversation. As it heats up dramatically, we start cutting alternatively from a close-up of one character (a), to a close-up of the other (b), and back to (a) again. The number of consecutive close-ups varies with the values of the scene: sometimes only two or three are used, but a conversation of some length or importance may require a good many more. The succession of close-ups may be interrupted; a group of five or six close-ups may be followed by a fuller shot for a change of rhythm, emphasis, or position, then the (a), (b), (a), (b) series may be resumed. Occasionally, a long series of close-ups will be sustained, but the practice is exceptional. At one time it was considered proper to arrange cuts in multiples of three, and in a very short scene that is still a common practice. But for an extended sequence that convention is now deemed archaic, and the number of close-ups is mandated solely by the dramatic needs of the scene.

*See Dmytryk, *On Film Editing* (Stoneham, MA: Focal Press, 1984).

As the analysis of the *Casablanca* excerpt showed, a close-up is rarely a true point of view shot (unless the characters are standing nose to nose, and then a very close profile two shot is more effective). Although each separate close-up in a "separation" series is technically the other participant's POV each functions at a much higher level when it is also the viewer's POV—in other words, when the viewer feels the character he watches is speaking to him as well, and that he is directly involved in the verbal exchange and the emotional development. To be sure, a great deal depends on the quality of the material and the personalities and talents of the actors, but the size of the shot and the focus of the actors' attention also plays an important part. The viewer's engagement is roughly proportional to the size of the close-up and the direction of the actor's "look." The most effective angle can be within 2 or 3 degrees of camera center, when the actor is looking almost into the viewer's eyes, although it is usually a little wide of that mark.

When used in dialogue scenes, the function of "separation" is far more valuable for its capacity to display *reactions* than for any possible enhancement of the dialogue itself. A spoken line usually delivers a clear message regardless of the size of the shot, but a well played (which usually means underplayed) reaction to the words needs to be caught and interpreted by the viewer. Once that is accomplished, the viewer anticipates the response and swings his attention to the other actor's reaction, and the process of participation is repeated. Given well-conceived and well-played dialogue, the filmmaker's most important concern is the enhancement of the viewer's involvement in the scene through his choice of set-ups and their optimum juxtaposition during the editing process. Because, at their best, films are still moving *pictures,* and in a very real sense their artistic importance rests in what the viewer *sees.* What he *hears* is underscoring, whether it consists of words or music, and a director who favors the dialogue over the reaction is manufacturing a play, not creating a film. He is also depriving his work of the one quality that makes his medium unique—the viewer's emotional participation.

The best examples of "separation" are those in which reaction spars with reaction, and dialogue is minimized. The following sequence from *The Young Lions* is one such example.

Christian Diestl (Marlon Brando), a lieutenant in the German

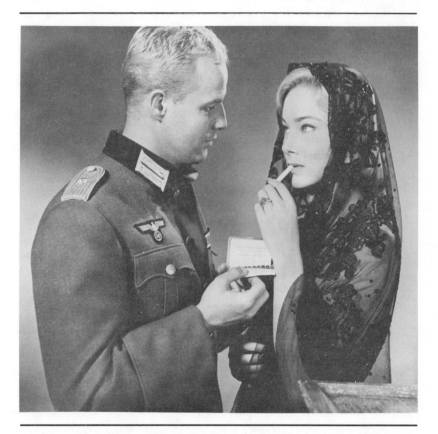

A seduction in silent close shots. Marlon Brando and May Britt in a scene from The Young Lions. *Photograph courtesy of 20th Century-Fox.*

army, is in Berlin on furlough. His superior, Captain Hardenburg
(Maximilian Schell) has asked him to deliver a present, a black lace
mantilla, to his wife, Gretchen (Mai Britt). Christian arrives at
Gretchen's apartment just as she is about to leave for an evening
out with a general. She obviously has an eye for an eligible young
patriot and she invites him to await her return. Besides, in wartime
Berlin available hotel rooms are impossible to find.

When she returns, he is lying on the floor, asleep and a little tight.
She asks for a glass of vodka, then opens her husband's package. She
moves to a mirror at the far end of the bar and, in a close-up, artfully
drapes the lace scarf over her blond hair. The excerpt starts with
this close-up.

	Shot	Length in seconds	Total
1.	C.U. Gretchen.	6½	6½
	She faces the mirror, drapes the lace scarf over her hair, looks provocatively demure.		
2.	C.U. Christian.	4	10½
	Still lying on the floor, he watches her, assessing the situation. After all, she *is* his captain's wife.		
3.	C.U. Gretchen.	9	19½
	She turns to face Christian, smiles slightly. Finally, she says, "Get me a cigarette, please."		
4.	C.U. Christian.	8	27½
	A slightly delayed reaction— "Huh?" She speaks, offscreen. "On the table." He looks around, spots the cigarettes on a table away from Gretchen. He nods and starts to get up.		
5.	FULL SHOT—Christian. Back to camera.	6	33½
	He rises, walks to the table, reaches down for the cigarettes.		

6.	C.U. Gretchen. She regards him steadily.	4	37½
7.	Tight O.S. Close-up Christian. He enters the shot, regards her with a tentative smile.	6	43½
8.	Tight O.S. Close-up Gretchen. Her look is less tentative than his. Slowly, she takes a cigarette, puts it in her mouth, reaches for the match box he holds up to her. Then, she scrapes a fingertip along his hand, as though she were striking a match.	15	58½
9.	Tight O.S. Close-up Christian. He looks at her appraisingly— snaps the match box shut.	5	63½
10.	Tight O.S. Close-up Gretchen. He raises his hand, gently removes the cigarette from her mouth, tosses it away. Then he reaches up with both hands and slowly removes the mantilla, dropping it onto her shoulders.	17	80½
11.	Tight O.S. Close-up Christian. Slowly, he starts to pull her toward him.	5	85½
12.	Tight O.S. Close-up Gretchen. She allows him to pull her to him; then, just as their lips are about to meet she quickly, teasingly, averts her face. Just as quickly, he straightens her out. They kiss, and the scene	5	90½

DISSOLVES TO:
An unrelated sequence in New York.

Throughout this sequence, which lasts 1½ minutes, there is little question about Gretchen's intent, but there is some doubt about

Christian's ultimate decision. The looks are long, the reactions subtle and the moves are slow, with one exception—Gretchen's close-up (6) considerably shortens the real time needed to get the cigarettes, the matches, and to cross the room to deliver them. Once Christian is on his way, in cut 5, the viewer is eager for the confrontation he has been anticipating.

The clear, though never broadly played, looks and reactions are quickly understood, but the viewer is allowed a good deal of time to savor them. The seduction is so involving and so complete that no scene of lovemaking is necessary or desirable. The close-ups of Christian and Gretchen have said it all, teasing and engaging the viewer as much as they have teased and engaged each other.

The "separation of characters" is probably the most common convention of present day filmmaking, and certainly one of the most valuable. It allows the editor to "play with time" (see Chapter 13), to accentuate reactions, to increase or shorten cuts to better accommodate the viewer. Incidentally, in skillful hands its idiosyncratic manipulation will inevitably work to establish the filmmaker's (or the editor's) style.

9

Rules and Rule Breaking

The masters of the screen have given the world a sizable stockpile of material containing beauty, entertainment, food for thought, and occasionally even a new slant on life, but they have not passed on any rules for duplicating their virtuosities. In a continuing effort to correct the oversight, film scholars have made it a practice to analyze selected sequences from classical films made by noted directors. Unfortunately, this leads only to "rules" for the sequences analyzed; rules laid down by the analysts, not by the films' creators.

By definition, creative workers in any field are not rule followers, but rule breakers. A dissection of, say, Eisenstein's "Odessa Steps" sequence shows us only its final version in the finished film. It tells us nothing of the hundreds of ideas created and discarded in pre-production, of scenes shot and never used, of countless changes in order and timing during the editing process. Nor can it assure us that out of that welter of material no better sequence could have been created. It only says this: These are the cuts and this is the timing that makes *this particular* sequence what it is. An inferior director, shooting a similar scene, might try to copy it, but a good director in the same situation would work doubly hard to avoid "borrowing" from the original creation.

The happy truth, at least for those who revere originality, is that there are few, if any, *specific* rules for creating a work of art. There are no esoteric formulas, no Phythagorean sets of sacred numbers. A unit of three may work wonders in one sequence while a unit of one, or eight, will best suit another. A sequence can be introduced with an establishing shot or with a giant close-up. Dialogue can, and should, be spoken at a faster pace than in real time, and a scene of violent and complex action often achieves its maximum effect when slowed down to a crawl.

It is also comforting to know that no two creative directors will make identical pictures out of a common script, nor will two creative film editors arrive at the same result when cutting identical rushes.

On the other hand. . . .

Suggestions can be made. Practices which have survived for more than half a century can claim the status of rules which should be understood and used, if only to build a solid takeoff platform for hoped-for creative flights.

The study of these rules or, more accurately, usages, can be made easier if the narrative film is divided into two broad categories— categories which sometimes overlap.* The first includes mystery, detective, suspense, and action films, in which the dramatized problems almost always involve murder, mayhem, or criminal activity, none of which, we like to believe, is exactly normal. The second takes in those films informally called "serious," or "adult" (in the nonpornographic sense); films which investigate and dramatize the more general problems of imperfect souls in an imperfect world, problems which are more damaging to psyches than to physiques.

A second category film with a solid story can be shot in a technically simple, straightforward style and still be impressive. The only real criteria of its worth are its honesty, its relevancy, and the believability of its characters. If, as is certainly desirable, technical adroitness is present, it serves only to *enrich* the impression, not to *make* it.

The second category is represented by films like *Sunrise, The Grand Illusion, Dr. Zhivago,* and *One Flew Over the Cuckoo's Nest.* Occasionally a director will cross the line, as Billy Wilder did with *Double Idemnity,* and occasionally a film noire like *Crossfire* will be equally at home in either category.

The sophisticated comedies of Lubitsch, McCarey, Sturges, Stevens, and a few others, are a subclass of the second category. And,

*The division into categories or subclasses is arbitrary. It is used only to simplify the problems of technical analysis and has no bearing on quality or capacity for entertainment.

of course, a number of films resist such primitive classification; they occupy various gray areas between one and two.

A first category film is almost always linear, with little or no subplot. Substance and character development usually play second and third fiddle to scenes of peril, startling confrontations, and related elements of mood and suspense. These aspects of the film are unquestionably enhanced by artful techniques, high-contrast lighting, and mind-boggling effects. When these elements abound, *they* become the message, and shortcomings of content and character development are cheerfully overlooked.

Films like *The Maltese Falcon, Gaslight, Murder My Sweet,* and *Jaws* typify the first category, as does almost any Hitchcock film, especially *Psycho* and that classic fusion of mystery and suspense, *The Thirty-Nine Steps.*

Because such films depend more on technique and formula than on substantive content, they are easier to write, easier to play, and easier to make. And because techniques are easier to dissect impartially than emotional stimulus and effect, they are usually accorded the greatest share of attention in any analysis of filmmaking.

Before leaping into specifics, a further distinction should be made, a distinction crucial to any discussion of the methodology of filmmaking. Because sensitivity and finesse are, relatively speaking, more important in the second category than in the first, a distinct difference in staging and camera use is indicated. In category two, where elaboration of the content is of prime importance, scenes are rehearsed and staged for optimum positioning and movement of the *actors,* then set-ups are selected which will best transfer the scenes to the screen. This procedure may be called "bringing the camera to the actor." In category one, however, most films depend on sequences of *effect* rather than of *argument,* which makes the visual presentation of the scene a major consideration and often imposes restraints on the actor's freedom of movement and relative positioning. Although, as always, scenes are rehearsed before the set-up is locked in, the staging is more likely to be arbitrarily manipulated by the director in the interests of melodramatic composition (category one) rather than dramatic reality (category two). This procedure may be called, "bringing the actor to the camera."

Although rehearsal and staging precede set-ups and lighting, it will be advantageous to consider the last item first. Lighting styles have had their ups and downs over the years, but certain conventions have been with us nearly as long as the art itself. Let us begin with the simplest, broad comedy, which utilizes full shots much more than close-ups. Comedians move a lot and rarely hit their marks, subtle beauty is not often a major consideration, and the desired mood is almost always upbeat. Even "black" comedy needs light to make it palatable. The recipe? Full, *flat* lighting, which covers the entire area of action. Lovers of comedy care little for high cheek bones or mood-inducing shadows when looking for laughs. Pictorial subtlety is not an advantage; the main consideration is pictorial clarity.

Clarity is also desirable in second category films but, at its best, it is clarity of depth, achieved with finesse. And it is expensive. In the extreme, every piece of furniture, every light fixture, every bit of bric-a-brac on the mantel is lit with loving care. The pipe on the piano receives its own key, side, and back lights. Everything is clear, but nothing is flat.

Even when modified the drawback of this style is its drain on time. Punctilious lighting on a closely-budgeted film is achieved at the expense of rehearsals, staging, and acting; in other words, at the cost of performance. No filmmaker should allocate more time and money to lighting a set than to improving the film's content and delineation. More than one film has been made in which the beauty of the sets and their occupants has overshadowed the impact of the message, but rarely has any of these charmed the viewer.

At the opposite pole is the lighting used in "film noir," and almost any first category picture. Indeed, in production circles the terms which describe such lighting, *low key*, and *high contrast*, are often used to define the category. Here, clarity is often an enemy. Shadows become "characters." Every dark corner harbors a threat, every half-lit face conceals a mystery, a facet of the character left out of the script. The viewer looks for the obscure clue, anticipates the imminent surprise, the hovering menace. For him the classically-lit scene is of little interest; here the suggestive outshines the definitive and is far more "realistic" than the real.

This style was pioneered by men like Murnau and Lang. It was buried with the advent of sound, then disinterred by Orson Welles. But it was its adaptability to the harsh demands of time that gave it its greatest impetus. In the early forties, ambitious B directors, working on extremely short schedules with inferior scripts and second rank casts adopted and expanded the technique to improve their

Rule breaking sometimes results in the creation of new categories of film. In this scene from Crossfire, *low-key lighting creates the unsettling mood characteristic of film noir. Robert Ryan and Robert Young are at right. Photograph courtesy of R.K.O. Radio Pictures, Inc.*

opportunities to advance to bigger films, fatter fees, and higher status. After all, one light and a cleverly placed cukullaris casting its unreadable shadows on a wall could sometimes seduce the imagination of even a very perceptive executive.

But, the fact that film noir lighting does not submit to the rules of classic "naturalism" stimulated experimentation at all levels and resulted in aesthetic and dramatic enrichment of the film medium. One of its greatest advantages is its liberation from any obligation to a realistic on-screen light source, an obligation that can severely limit creative lighting. As Matisse said, "Exactitude is not the truth."

In the film *Crossfire*, a scantily furnished, high-ceilinged foyer in a seedy apartment house features a flight of stairs built along one of its walls. A room on the second floor landing, adjacent to the top of the stairs, is the setting for a murder and the eventual entrapment of the murderer. The stairway is used for several entrances and two or three short, suspenseful scenes, and it is essential that these ostensibly minor set-ups establish the mood for what follows inside the room.

Every lobby shot includes the stairway, either as a setting for an entrance, a short confrontation, or simply as a background for a scene on the second floor landing. The set was lit with a single arc light placed on the lobby floor some distance from the stairs. The resulting larger-than-life shadows of the stairway's railings were thrown across the steps and onto the wall, each slanting to a slightly increasing degree on either side of the shot's center. The strong contrast of light and shade, the clear symbolism of the vertical railings and their oversized shadows through which the scene's characters moved, provided an arresting setting—eerie, but completely believable—as long as the viewer was not allowed to study it too closely.

Over a period of years those scenes have been shown to hundreds of film students. Not one has ever challenged the reality of the offscreen light source, or realized that it was placed in an unrealistic, even impossible, location. The success of such deception depends on controlled distraction, on "managing" the viewer's attention. This is accomplished by making sure that every frame of every scene is relevant to the plot, that the viewer's involvement *in the film's characters* is sustained, that the set functions only as a setting, an aura, an ambience which enhances the intent of the scene without attracting attention to itself. The unreal, or outré, aspects of lighting in any scene must be used to accentuate the "essential reality" of

A fine example of creative lighting. Arthur Franz and Arthur Kennedy in a scene from Anzio, *a DiLaurentis film.*

the subject, never as a spectacle in itself. That way leads to bad taste, bad films, and bad art.

It must be clearly understood that the realization of the lighting under discussion, in any category, is the responsibility of the cinematographer, not the director. The director must have a concept of the mood he wishes the film to project, and he must be able to communicate that concept to his cameraman; then he must go about his own business while the lighting man creates. Given talent, and the freedom to exercise it, his creations will probably exceed the director's fondest hopes.

10

The Modification of Reality

A well-known Hollywood folk tale would have us believe that film-makers once jotted down their concepts on the backs of menus while dining at the Brown Derby. If, as sometimes happened, a director's ideas flowed too freely, his scrawls could cost him the price of a table cloth. The tale is probably true. It was easier then to create motion pictures; "silent" ideas could be doodled in fragmentary fashion and later fleshed out in action on the set.

Then came sound, and words, and literary practices and limita-tions. Now many filmmakers are not even aware of the unique possibilities of their medium as they struggle to manufacture screen drama with verbal rather than visual signs, with shots that record rather than reveal.

A semantic axiom states that *words* are signs for *things*, but only in a generic sense. To make things more specific words must be modified by *other* words, and it often requires a paragraph of mod-ifiers to pin a word down to a sign of a *particular* thing. *Shots*, on the other hand, are by their nature already signs for particular things— a shot of an airliner, for instance, will identify it specifically as, say, an American Airlines 747 rather than just an "airplane"—and are therefore much more real than words, semantically speaking. Using the bare shot descriptions of Chapter 3, shots do picture reality; the camera is at average eye level, the lens shows a normal vanishing point, and the lighting is impartial. Utter reality!

But, as Henry Moore said, "Merely to copy nature is no better than copying anything else," and nothing in art is quite as dull as honest-to-goodness reality. The truth is that no good film shows reality as we ordinarily know it; it shows only a particular film-maker's perception of reality—truth as distilled by an opinionated storyteller with a gift for modifying the "spitting image."

So, while words are modified to make possible a more accurate description of a specific thing, shots are modified to facilitate the realization of a more indeterminate, a more universal, reality. Al-though the modifications seem to work in opposite directions, they serve the same purpose: to more deeply engage the viewer's (reader's) interest and involvement. And, coincidentally, it is largely through creative use of shot modifiers that a filmmaker exhibits a personal technical style.

Such modifications which, ideally, should result in cinematic scenes that would be difficult, if not impossible, to achieve with words alone, are implemented through lighting, lenses, and camera positioning.

It is no secret that in no more time than it takes to set a few lights, a good photographer can change an ugly duckling into a swan— a magical feat accomplished nearly every day. In John Ford's *The Informer*, clever lighting and a shawl transformed a saint into a prostitute. And many an unscrupulous news photographer has given a frightened and possibly innocent suspect the appearance of a man-iacal killer by cunning manipulation of his photoflash. Similar mod-ification can be achieved in an inanimate object—any object. Converting the appearance of a warm, comfortable Victorian resi-dence into Poe's House of Usher is only a trick of the trade.

An actual third dimension is the only ingredient of reality usually missing in a film. "Left" and "right" are omnipresent; "up" and "down" only a little less so. But "to" and "fro" is an optical illusion which, in spite of the fact that a sense of depth can lend more "life" to a scene than any other technical factor, is completely absent in most films. Part of the problem is *color*.

Properly used, color has been a decided asset on the screen, but it is not an unmixed blessing. Color film is more forgiving and less demanding than black and white, and just as the automobile has contributed to the softening of many a leg muscle, color has eased the demands on the creative intensity of many a photographer. For example, in black and white certain shades of red and blue register the same shade of gray, and they must be "separated" by lighting and by staging. Color *inherently* separates any and all hues and, for the not-so-great photographer, flat lighting has become the easy way out. But flat lighting is also featureless lighting. An impression of the third dimension can only be achieved by the creation of a series of planes in depth through the proper juxtaposition of light and shadow. This is true not only of the fuller shots but of close-ups as well.

A character lit "in depth," showing strong shadows and differentiated facial planes, will certainly look more interesting and attractive—yes, even "deep"—than he or she would in a flatly lit magazine style shot.

The filmmaker will ask for the lighting moods and effects he wants, but it is the cinematographer who must deliver them. Since there are probably more cameramen equal to the task than there are directors who know what to ask for, this rarely presents a problem. However, the choice of lens and camera position is a different kettle of fish; it always belongs to the director. In both of these areas the filmmaker must first know exactly what effects, impressions, or impacts he wishes to achieve, and second, what lens and set-up will best accomplish his desires. These two factors further modify the lighting effects referred to above, and the reality of the scene.

The 50mm lens is usually the workhorse of the film; it is the camera's *normal* eye. It shows the shot's subject at about the same distance and the same size as when seen with the human eye. The compositional perspective and the relative distances of other objects or people in the shot, whether in front of or behind the central subject, also appear to be normal. But, the 50mm aside, all other lenses distort the image to some degree and offer innumerable opportunities to cheat reality.

The *narrow angle*, or *telescopic*, lens brings objects closer than normal; the extent of the magnification is a function of its focal length. A 75mm, for instance, decreases the apparent distance from object to viewer by approximately 50 percent, while the 500 is truly telescopic. Outside of their obvious but seldom used advantage of facilitating close shots of distant objects, the narrow angle lenses are used chiefly to enhance appearance. Since there is less depth of focus and a flattening of features, the 75 or 100mm lens obtains

The director must construct his or her vision of the film's narrative by determining each set-up. The author checks a set-up with cinematographer Joe MacDonald. Photograph courtesy of Columbia Pictures Corporation.

more flattering close-ups of women. The focus can be concentrated on the eyes, leaving everything else, before and behind, slightly, though never obviously, indistinct. And even though the background is proportionately closer to the subject, the shallow focus renders it somewhat obscure and allows the subject to stand out from the surroundings.

The foreshortening property of the narrow angle lens serves to create a type of movement which is all the more effective because of its infrequent use. For example, shot with a 500mm lens a sprinter, running full tilt at the camera, seems to be taking steps so short as to be nearly running in place. So much effort expended in making so little progress creates a heart-pounding effect quite like that in the nightmare in which the dreamer strains every muscle yet finds himself unable to escape an ambling pursuer. For this and similar purposes this technique is far superior to the floating slow motion shots so frequently used, though the two modes can sometimes be combined to advantage (see page 122).

The wide angle is a much handier lens; it also requires more knowledgeable lighting, since it deals with the indepth problems previously mentioned. And it is the only lens that can convey a sense of the third dimension. In short, the wide angle lens appears to push the subject away from the viewer. A 25mm lens, for instance, places an actor at what appears to be double his real distance from the camera, while the distances of background objects or people increase in the same ratio. A very important aspect of this property is that a person walking away from or toward the camera appears to be covering *twice* the normal distance with each normal step, or would if the viewer's points of reference for judging distance and speed of movement were not equally illusory.* But they are, and the distorted images are accepted as normal. The strong subliminal effect, however, offers opportunities for distinct aesthetic enhance-

*The use of space-distorting lenses is most graphically illustrated in TV commercials. When a car manufacturer wishes to demonstrate the maneuverability of his product, a narrow lens is used to shoot down a long line of red cones. As seen on the screen, the cones seem much closer together than they are in actuality, and the car manages to dodge them quite nimbly as it zigzags down the line. On the other hand, when the selling pitch is rapid acceleration, they shoot down a road with a wide angle lens. The car then seems to devour a quarter mile of road in the time it takes an ordinary car to travel two hundred yards—because that is the distance the filmed car really covers. In both cases, what you see is *not* what you get. In a film, however, it may be what you're looking for.

ment of many scenes in the area of movement through the application of *dynamic* composition, whose rules diverge widely from those applied to static groupings, and arbitrarily disproportionate physical relationships can be filmed to advantage.

One obvious example is a common movement in a ballet. A dancer glides rapidly from the distant edge of the stage. A few graceful steps and he executes a flying leap. In this common maneuver one dancer excels another only because he can run a touch faster, leap a few inches higher, and possibly do both with more finesse. But these slight advantages rarely add up to an awe-inspiring difference.

On the screen, however, even a routine dancer can be presented as a truly superhuman performer. Shot with a low camera (perhaps a foot off the ground) equipped with a 25mm lens, his few running steps traverse a remarkably wide stage (double the actual width) with apparent ease. (Since the length and speed of his steps will also be exaggerated by the 25mm lens, he can afford to take shorter strides and keep them under better control than they might be for a "live" performance.) He moves, of course, *toward* the camera (unlike a theatrical viewer, the camera will be on the stage) so his figure increases dynamically in size until he leaves the floor in a prodigious leap and soars over the camera (the viewer) in truly breathtaking fashion.

All the separate aspects of this maneuver—the run, the leap, the direction of his movement, and his spatial relationship to the film viewer—have been aesthetically far more exciting and engaging because, although the camera has photographed the action, it has *not* been used to "promote the redemption of physical reality." On the contrary, it has intentionally distorted speed, height, distance, and dynamic growth to promote a deception which the viewer will enthusiastically accept as reality, and which he will long remember, but which, in the real world, he will never be able to find.

Close, motionless compositions can also profit from the use of a 35 or 25mm lens. The features are subtly distorted; the nose is longer, the eyes set deeper, bone structure and skin imperfections are more sharply defined. These factors can be used to bring a strong, down-to-earth character to the screen or, when pushed to excess, one who is physically or morally somewhat less than attractive. And the fact that they are undefinably offbeat helps to make the screen characters more interesting to the viewer. Though rarely mentioned by critics or analysts, wide angle lenses are essential to any production with film noir pretensions, and to "horror" and "who-done-it" films as well. (In my own work, the 40mm lens, because of the slight subliminal distortion which I preferred, was my "normal" lens in any genre.)

With the exception of the 50mm, every lens distorts reality to some degree, but the distortions should rarely be noticeable. When the degree of *distortion* is properly coordinated with the *character(s)* being photographed, the effect should be subliminal, and it will be accepted by the viewer as an integral physical singularity of the person he sees on the screen. The discriminating use of the lens system can take some of the burden of characterization off the actor's back and allow him to "get into" his screen persona in the most honest way possible. When properly understood and used, this is one of the great advantages of the film medium. (This area will be more fully discussed in Chapter 12.)

The third shot modifier, *camera positioning*, is dictated partly by the set-ups, but more positively by the taste and inclinations of the filmmaker. Ordinarily, the eye level shot adds nothing to the screen—it is neutral and bland—and since few viewers look for neutrality in a film, it should be avoided. Of course, certain camera positions are mandated by the scene's personnel. For instance, in a dialogue scene involving James Garner and Sally Field, over-shoulder shots and close-ups would angle *up* at Garner and *down* at Ms. Field, and if a man on the ground chats with the fiddler on the roof, the same camera positioning would be called for, though angled to a more extreme degree. But the greater part of any film is on the level, cinematically speaking.

Arbitrarily setting the camera off eye level has many advantages. Perhaps the chief one is similar to that engendered by lens distortion—it puts the scene or the players slightly out of kilter. It is a consciously imperceptible nudge to the viewer's sense of perception, and it helps to keep him mentally on his toes. In most set-ups the preferred position is *below* eye level, where it serves a practical as well as an aesthetic purpose. People look down on the ground they walk on more often than they look up at the stars they wish on. A person may look skyward when daydreaming, but daydreaming is rarely the subject of a scene. As Rodin understood so well, thinking lowers the head, and a person usually looks down when speaking on the telephone or kicking a clod, for example. From a normal or slightly above normal point of view, the camera looks at eyelids

Sometimes the close-up says it all. Fred MacMurray in The Caine
Mutiny, *a Columbia Pictures film.*

that screen off the eyes and conceal the actor's thoughts, but a shot which looks up at a film's character serves not only to jog the viewer's awareness but gives him a clear view of the player's eye reactions in the bargain.

In long shots, and especially in crowd shots, the preferred camera position is *above* eye level. At eye level, increasing distance shows more and more "head room"; that is, space above the action. With some exceptions that space usually says nothing and makes a very bad composition.

As for the degree of tilt, the *up* tilt may vary from 3 to 6 inches in a close-up to a foot or more in longer shots, but in all circumstances, caution should be observed. The upward tilt has become a common practice which, too often, is merely common. However, the technique is too useful to abandon; it should just be approached with special care. A shot tilted *sideways* is rarely of any value in straight narration and should be avoided, but the *down* tilt is often useful and, in its variations, offers a great deal of latitude; in extreme long shots the camera is often positioned many yards above eye level. The exact distance varies with the situation and, once more, with the filmmaker's instincts and taste. And as in every aspect of the art, there are frequent exceptions to the rule.

Just as authors enhance their writing by their individualized use of word modifiers, so a filmmaker "stamps" a film through the personalized choice of shot modifiers. The subtle use of this facet of filmmaking is of extreme importance both to the filmmaker and to the aesthetic singularity of his work.

11

Symbols, Metaphors, and Messages

The Chinese character 女 means *woman*. Now that you know that you will recognize it anywhere, without regard to what other languages or dialects may name it. You do not have to know its Chinese word, or how the sound of that word transliterates into any other language. 女 means *femme* to a Frenchman, *mujer* to a Spaniard, and so on through all the world's tongues. That is the beauty and the advantage of a symbol as opposed to a word. That is also the beauty and the advantage of a film—at its cinematic best.

Like any animal, humans communicate with sounds (words) and gestures (images). But, at least for humans, symbols do not eliminate words, they merely transfer the responsibility for their interpretation. We automatically translate gestures into more words. A "wave-off," for instance, can be confusing until the observer translates it verbally as, "Go away!" For, just as an idea has no meaning unless it can be described in words, so an image makes no sense unless it serves as the bearer of a verbally describable emotion or bit of information. To cinematic purists this is one of film's most troublesome contradictions. But when we speak of cinematic imagery what are we really talking about? We're talking about the central element of a three-step operation.

It always starts with words—words searching for a viable theme, words identifying the growing elements of a story, and finally the words that bring the crystallized concept to life. Unfortunately, this is also where it most often stops. In a novel, a poem, or a play, the words are all, or nearly all, of the realized creation, but the filmmaker knows that this is only the *first step* of a structural sequence. That is why, except for purposes of promotion, the original treatment or script has no need for beauty of expression, only clarity—clarity of concept and of purpose.

The *second step* is the transposition of the original concept into images, a step usually made most effectively by the filmmaker. The images which communicate and dramatize the concept should be a carefully selected as are the words of any master writer, be it Poe or Proust, because in the *third step* they must deliver the original message, but usually not in the original words.

The purpose of the second step is to stimulate the viewer into *his own* interpretation, at *his own* level, in *his own* words, of the filmmaker's concept. It is this singular characteristic which has made motion pictures the most universally popular artistic medium in the history of the world.

Perhaps the most abstract, yet completely understandable cinematic sequence ever made is Hitchcock's "murder in the shower" montage in *Psycho*. During nearly one hundred cuts, all short, many undecipherable, and none explicitly showing a knife entering the victim's body, the verbal message is, "My God! She's being killed in the bathtub!" And the montage allows each viewer, thinking in his own language and using as many of his own words as he cares to use, to deliver the verbal message to himself. Let us consider a few other examples, from simple to abstract.

In *Murder My Sweet*, Philip Marlowe (Dick Powell) and Moose Mulloy (Mike Mazurki) leave Marlow's office to visit a night club. As they exit the room, the camera swings into a close shot of a water cooler, surmounted by a five gallon bottle of water. A large air bubble "blubs" to the water's surface, and the scene dissolves into the night club's neon sign, FLORIAN's. Two of its letters flicker fitfully, and the viewer's mind immediately registers, "A dive!" He not only knows what kind of a joint he is about to enter, he also begins to anticipate its possible dangers.

All this information is not delivered by the sign alone, but by a layer of images. It is introduced by the shot of the bottled water. (The image, filmed in 1944, has since been borrowed for a few other films.) Accompanied by its distinctive sound, the shot serves several purposes. First, it always elicits a laugh, a welcome break in a series of tense sequences; it furnishes a short breathing space while holding the viewer's attention; it provides both an audible and a visual "springboard" into the next scene. Here the image of the amorphous bubble is augmented by the shot of the sputtering neon, and its vulgar "burp" enforces the tawdriness of the sign and the enterprise it advertises. There is no need to further "establish" the character of Florian's Bar—we can proceed with the plot.

In *The Caine Mutiny*, shortly after taking command of the ship, Captain Queeg (Humphrey Bogart), while talking to his junior officers, is faced with a troublesome situation—the shirttail of one of the ship's crew hangs outside his pants. Queeg, a martinet, frowns, then his right hand slips slowly into his right pocket. Is he reaching for a pack of cigarettes? At this point, both the attendant officers and the viewer can only guess. But when Queeg's hand reappears it holds two steel balls. Spasmodically, his fingers knead them in the palm of his hand, and only their metallic clicking is heard as the junior officers watch in disbelief.

The viewer's first reaction is laughter, but he quickly realizes that for Queeg this is a crisis, and that the steel balls are his security blanket. From then on throughout the film, whenever Queeg's hand slips into his right hand pocket, the viewer knows that the Captain is suffering an attack of paranoia, and he reacts accordingly.

Until the film's climax the routine of the steel balls is a device, a sure source of humor which firmly establishes Queeg's problem, but at the mutineers' courtmartial it plays a leading role. As Queeg testifies to the events leading up to the "mutiny" he maintains the calm demeanor of a well-balanced man and officer, smoothly refuting the junior officers' testimony. Then the defense attorney catches him in a lie. Queeg stammers, his hand slips into his pocket, brings out the steel balls, and the viewer knows that the Captain is finished, that he will expose himself and his paranoia. No words are needed to bring home the point and none could possibly do so in such a remarkably short time.

This excellent cinematic device was taken in its entirety from Wouk's novel, but it truly reached its fulfillment in the film, where the viewer could see it all *happening* instead of taking someone's *word* for it.

Another scene which lends itself to imagery was written by John

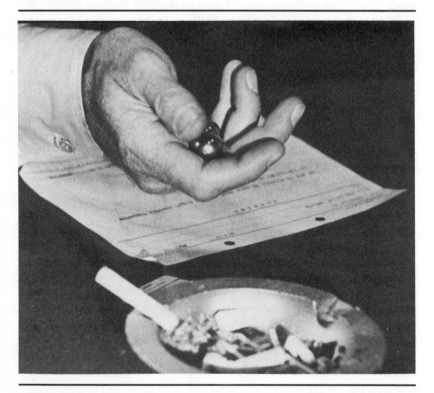

The perfect metaphor—Humphrey Bogart's steel balls in The Caine Mutiny.

Fante in *Full of Life*. Here a young hollywood writer, a second generation Italian, returns to his family home in Northern California to fetch his father back to Los Angeles for the birth of his first grandchild.

The place is typical of many rural homes; the spacious yard is littered with rusty odds and ends and used-up pieces of ancient equipment, in the midst of which, prostrate on a worn-out settee, the father is taking his afternoon nap. The writer makes his way through the clutter to his father's side and looks down at the old man—at his wrinkled features, his gnarled hands, his used-up body which seems too small for the work clothes he wears—and his eyes fill with tears as he remembers the powerful stonemason he knew when he was a boy.

A fly drones lazily in the humid summer air. And suddenly, like a striking snake, the father's hand darts out and snatches the insect in mid-flight. The image of the lightning move instantly transforms the pitifully weak old man into a forceful patriarch, and establishes a character that will remain consistent throughout the story. The tale is well-written—but *seeing* is *believing*.

In *Crossfire*, a young man narrates his movements of the past evening. He has left an apartment and gone out into the street for a breath of sobering air. As he talks, a subjective camera, its fuzzily-focused lens standing in for his alcohol-befogged eyes, moves down the sidewalk recording an undecipherable street sign and some shadowy, shuttered store fronts. At the end of a short, aimless cruise, it swings over to the curb and focuses in on a large metal trash can just as a jazz trumpet smears out a harsh opening blast. Simultaneously with the sight and sound, a quick, short dissolve discloses a huge close-up of a smiling Ginny. No words are needed.

The viewer understands her character at once, and he supplies his own epithet, which is in no way abstract. The subsequent scenes fill out and modify the original impression, but the image of the garbage can and the sound of the blaring trumpet have given him a great deal to build on—all in a matter of seconds. No verbal description could have been as concise, as colorful, as universally understandable, yet as specific in revealing appearance and *character* as these three or four seconds of carefully abstracted metaphor.

However, speed is always a relative consideration. The viewer's response to the scene's recommendation is not always immediate,

and sometimes he must be allowed an extra moment to work out
the relevance of his own experience to the image's implication.

In *The Verdict*, Frank Galvin (Paul Newman) is trying what may
be his last case in a Boston court, a case of medical malpractice. The
defendants—a hospital and two doctors—are playing with a stacked
deck. They have the best and the slickest lawyers, a friendly judge,
an attractive "mole" ostensibly helping Galvin, and the advantage
of disappearing witnesses. It is not surprising that Galvin has lost
all hope. But in a last minute effort he succeeds in locating the nurse
who admitted the victim for the operation. Because she can deliver
evidence supporting the patient's cause, she has been threatened,
forced to leave Boston, and to give up nursing. She now works as a
teacher in a New York nursery school.

Fortuitously, the trial extends beyond the weekend. Galvin flies
to New York and finds the ex-nurse, Kaitlin Costello (Lindsay Crouse),
playing with a group of children in the school playground. He pre-
tends to be casing the place on behalf of his 4 year old nephew, and
exchanges a line or two of small talk. Then, in a full shot, across
Kaitlin and the children, with Galvin in the background, he says:

> GALVIN
> Oh, you're the - you're the one they said was a
> nurse.

> KAITLIN
> Who told you that?

> GALVIN
> Oh, I don't know - uh - Mrs. . . .

CLOSE GROUP SHOT   KAITLIN AND CHILDREN

During this cut the camera dollies in to a CLOSE SHOT
of Kaitlin.

> KAITLIN
> (helping out)
> Mrs. Simpson?

> GALVIN'S VOICE (off-screen)
> Yeah.

                              KAITLIN
                          (after a beat)
                I used to be a nurse.

                              GALVIN'S VOICE (o.s.)
                It's a wonderful profession. My daughter-in-
                law's a. . . .
                          (he pauses)

CLOSE SHOT   GALVIN
                              GALVIN (cont.)
                What'd you do - just stop?

CLOSE-UP   KAITLIN

                              KAITLIN
                          (another beat)
                Yes.
                          (she starts to turn away)

CLOSE-UP   GALVIN

                              GALVIN
                Why'd you do that?

CLOSE-UP   KAITLIN

She turns back toward him - is about to answer -
stops as she sees something.

CLOSE INSERT

A New York to Boston shuttle ticket envelope sticks
out of the breast pocket of his overcoat.

BIG CLOSE-UP   KAITLIN

She looks up at Galvin - a long, long look as it all
begins to sink in.

BIG CLOSE-UP   GALVIN

He moves up into a CHOKER - finally;

GALVIN
Will you help me?

BIG CLOSE-UP   KAITLIN

Near tears, she looks at Galvin. Finally the scene;

DISSOLVES OUT:

This is truly a cinematic sequence in the modern sense—a sequence of *mental* movement. Both minds are working and the viewer's mind works with them. Only the final question deals directly with the substance of the scene, but the viewers, who fully understand the lawyer's need and the nurse's predicament, follow the working of her mind as she silently deciphers the implications of the shuttle ticket. The viewers empathize with both characters, and though the nurse's decision is withheld in the scene (it would be counter-dramatic at this point) they know she will show up in court, and they look forward with hope and apprehension to that appearance. The filmmakers have supplied the background and the need; now they permit the viewers to live through the crucial scene with the people on the screen. Like all good drama it is subtle, suspenseful, and inevitable.

In all these examples the viewers are allowed, encouraged, to supply their own interpretations and descriptions, in their own words, of what they see. They need not know the definition of paranoia, or even the word itself, to recognize the basis of Queeg's behavior; they need never have met Ginny to instantly have more of an inkling of her character; and they need never have been to Boston or New York to recognize the complete meaning, in this particular situation, of an air shuttle ticket.

Spoon-feeding viewers to make sure they get your, and only your, version of the message results in mediocrity. It may require some suppression of the ego, but creating a scene that offers the viewer a sense of discovery and involvement is certainly most desirable.

Finally, another of the filmmaker's collaborators who can contribute creatively to the "auteur's" benefit is the *actor*.

In most successful motion pictures, the director seeks the creative contributions of the cast. Here, on the set of **Raintree County,** *an M.G.M. film, the author discusses a scene with members of the cast, including Elizabeth Taylor and Montgomery Clift.*

12

Auteurs, Actors, and Metaphors

The auteur theory, as defined by Andrew Sarris, is applicable to filmmaking practice. However, like most theories of art forms (e.g., Stanislavsky's on acting) it has bred ridiculous and dangerous excesses. Many young directors now use it as a justification for self-indulgent, dictatorial attitudes, especially when working with actors, just as some insecure actors use a distortion of Stanislavsky's "method" as a justification for self-indulgent ego trips when working with their fellows.

The current truth is that *good* filmmakers, with rare exceptions, reject the concept of *sole creativity*; while retaining control, which eliminates the manifest shortcomings of corporate thinking, they welcome the actors' creative contributions. John Huston has said, "In a given scene I have an idea what *should* happen, but I don't tell the actors. Instead, I tell them to go ahead and do it. Sometimes they do it better. Sometimes they do something accidentally which is effective and true. I jump on the accident." He might have added that not infrequently the "effective and true" contributions are *not* accidental, but the results of original thinking. And while such contributions carry less weight in those first category films which depend largely on camera effects and technical contrivance, they are vital to the optimum realization of the characters in any film of the second category.

All scripts suffer from an inherent deficiency which is largely ignored or, unhappily, deemed an advantage; a film's characters are first developed by the writer, and they inevitably reflect some aspect of the writer's persona. It is impossible to create a group of completely authentic *individuals*—good, bad, or in between—when each is fashioned in the author's image, is an echo, however faint or distorted, of some facet of the author's being. Such characters are clonal versions of a single mind rather than the unique personalities they would be in reality.

The problem of character construction is further compromised when the film reaches the shooting stage. Auteurs who insist that characters be shaped only on the basis of the auteurs' interpretations of personal or vicarious experiences surely diminish the range and richness of their work. The film's characters can be fully realized only if the filmmaker takes advantage of the creative talents of those who must ultimately *be* those characters—the actors in the film.

The writer's script and the director's guidance can qualify the actor's interpretation, but they should not confine his intuition. The truth is that a good script is seldom ambiguous, and there are usually few differences of interpretation between writer, director, and actor concerning the character's broader aspects. But in the search for inner nuances and points of view which can be transformed into pertinent mannerisms and individualized attitudes and reactions, each actor, given the freedom to create, will add his unique persona to his film character. The director who encourages such creativity will find himself in command of as many fundamentally *different* characters as there are actors. On the other hand, the player who apes, or is forced to ape, the director's "vision" is rarely at ease with his screen character, and the camera's impartial eye will faithfully record his discomfort and pretense to the detriment of his performance and the film.

None of the foregoing should be interpreted as an abandonment of the director's responsibility. He is always the "yea or nay sayer," the amplifier, the refiner, the particularizer; the final decisions are always his. If he cedes some creative initiative in the area of performance, he does so in the interest of greater character depth and believability.

Although it is to be expected that filmmakers should differ in

On location in the French Alps, the author and Spencer Tracy discuss
a scene from The Mountain, a Paramount Pictures film.

their approaches to directing, to script "doctoring," and to working with actors, the rather surprising truth is that, broadly speaking, their approaches to the technical methodology of filmmaking are remarkably similar. Huston's method was neither original nor uncommon, and the more secure the filmmaker the more willing he is to consider an examination of a player's point of view.

On their part, experienced screen actors know that they, too, must surrender something to the director; they must give up "acting" as they learned it in drama school or in the theater. *Restraint* is the key to superior film performance, and knowledgeable actors are aware that directors have at their disposal a number of tools and techniques that help them to realize less "managed," more honest, more subtly defined human beings by freeing them from the necessity to strain their voices or to exaggerate their physical movements and facial expressions as they labor to delineate their roles.

The manipulation of lighting, lenses, lens filters, and camera movement can sharpen or diffuse the texture of the actor's skin, emphasize or soften the angularity of her features, lend menace or compassion to his gaze, and relieve him of the burden of excessive or unnecessary reaction. For instance, a scene requires the actor to maintain a poker face while receiving some shattering information, yet the viewer must realize that the apparent absence of emotion is not due to a lack of awareness or sensitivity. A very slow dolly or zoom-in from close shot to close-up will convey the sense of heightening, of the building of inner emotion, even though the actor's face and eyes show no reaction at all.

Sound manipulation can magnify a murmur into a roar or enable an actor's honest whisper to reach the viewer. And cutting can accentuate, even create, changes in physical pace or rhythm, thus eliminating the necessity for exaggerated or artificial movement, and furnish the surgical means for excising imperfect performances or undesirable appearances. In short, the actor does not have to "act" the character; he can "be" it normally and let the skill of the director and the crew dramatize it as they will.

It could almost be said that the film was born to glorify the metaphor, and under the constraints of the censor filmmakers once created metaphors in plenty. But most modern directors, with their proclivity for reality and the explicit, are not even aware of the potential of this remarkable artifice. A pity! Aside from relieving the actors of some difficult and talky scenes, as well as occasional

If the set-up is good, the best shot in the arsenal can be one of no reaction at all. Close-up of Jack Lemmon in Missing, *a Universal Pictures/Polygram Pictures film.*

pain, the metaphor can eliminate the director's struggle to pictorialize some cliché situations. It is difficult to describe the advantages to be gained from the creation of a few well-placed metaphors, but two rather simple scenes from *Mirage* will elaborate the point.

At the film's opening, a man (Walter Abel) falls from a twenty-seventh story window of a New York skyscraper. Gregory Peck witnesses the fall from the same window, but the shock of the incident induces temporary amnesia, and he loses all memory of the fall and the events leading to it. Throughout the rest of the film occasional short flashbacks help Peck regain his memory and reconstruct the tragic situation.

A few minutes after the fall, Peck enters a local bar where the patrons are discussing the "accident." As he passes a group of imbibers he hears one say that he had once dropped a watermelon from a fourteenth floor window to hear how it sounded when it hit the ground. "All right, how *did* it sound?" asks his drinking companion. "About the way *he* did," is the answer.

The viewer, along with Peck, hears only the dialogue, since the speakers are not shown. The line is laughed off as a tasteless comment and forgotten. But a few minutes later, the sight of a "flash" headline concerning the victim's apparent suicide triggers a brief flashback in which Peck "sees" Abel fall some distance toward the street. A direct cut then shows the street some 15 or 20 feet below the camera, and a large watermelon plummets into the shot and splatters on the sidewalk.

Certainly no stunt man could effectively duplicate the scene, and a dummy in such a situation is completely inadequate. But the watermelon *really* shatters, dramatizing the effects of a human body hitting the pavement without the negative aspects of gruesomeness. For most viewers it was superior to any possible "real thing," since it was surprising and shocking, but not repelling. They could see, hear, and accept without "turning off." In other words, they could remain involved in the film.

A short time later, Peck is confronted in his apartment by a gunman. During the ensuing conversation Peck surprises the intruder and, in a short fight, renders him unconscious. But we "see" this without seeing it. The scene presented two problems. First, Peck had a chronic back injury which prevented him from undertaking such action, and doubles, in the close quarters imposed by the small apartment, would have been an inadequate substitute. Second, shooting a fight is one of the most troublesome tasks in filming; the lode has been overworked. But another metaphor solved the problem while adding positive values to the scene.

Early in the sequence, the gunman had switched the TV on to a wrestling match since, as he put it, in this deeply "psycho" era it was the only game remaining where "you can tell who the good guys and the bad guys are." Now, at the moment that Peck throws himself at the gunman, a cut to the TV set shows the violent climax of the match, culminating in a fall. As the loser is counted out, a cut back shows Peck picking up the unconscious gunman's legs and dragging him out of the room. The metaphor was a complete surprise, but amusing and welcome, evoking a laugh from viewers and showing some violent action which, where most people are concerned, speaks no more of reality than did the smashed watermelon. And at this point in the film it supplied the obligatory story development.

There are, of course, opportunities to develop more profound and sophisticated metaphors, but that will never happen unless the writer and the director always keep the possibility of their aesthetic and practical usefulness in mind.

13

Time and Illusion

Albert Einstein wrote, "For us convinced physicists the distinction between past, present, and future is an illusion, although a persistent one." Mathematically speaking, time can move forward or backward with equal ease and propriety. In our "real" world, however, time plunges ahead relentlessly, and practical men have invented a vast assortment of timekeepers, from sundials and hourglasses to the most precise atomic chronometers, to record time's metronomic progress. But the more accurate they are as machines the less they accord with human experience, and human experience is the main ingredient of good films, which is why "temporal realism" is such a meaningless phrase.

Shakespeare wrote, "Time travels in divers paces with divers persons. I'll tell you who Time ambles withal, who Time trots withal, who Time gallops withal, and who he stands still withal."*

In reality time is always experienced subjectively. For the fallen fighter the referee's ten count "gallops withal," while it seems to stand still for his anxiously hovering opponent. Next Christmas is a quarter of a lifetime in the future for a child of four, while it bears down with indecent haste on the septuagenarian. A prisoner "builds time" with Sundays, laundry changes, and full moons—clocks mean nothing to him. But even in jail, time runs an erratic course; the last week before freedom moves at a somnambulistic snail's pace, which on occasion is literally too long to take.

*From William Shakespeare, *As You Like It*, act III, line 328.

115

A thousand examples crowd the mind, but the point is that time's pace varies with the person and, to go behind Shakespeare's words, *with the situation.* The filmmaker who ignores this fact is not dealing fully with his story or with his characters, for he should be aware that no other medium can approach his in analyzing and dramatizing the capriciousness of time. To further complicate the problem there is also the viewer factor. Since this is the easiest knot to unravel it will be addressed first, and a bit of film history will serve to simplify the explanation.

During the "quiet" era, film was exposed at the rate of 60 feet per minute (FPM), but few people knew then, or know now, that it was projected at about 72 FPM.* (In the last few years before sound both speeds were upped a little.) The viewers were quite unaware of the increase in the actors' speed of movement and accepted it as normal; as far as is known no director attempted to discover why this was so. The question became moot with the advent of sound, which required a fixed projection speed. Picture and sound track, each occupying a separate portion of the same strip of film, were shot and projected at the rate of 90 FPM. They still are.

But something had changed. "Temporal realism" had asserted itself. The images on the screen now moved at the same pace that the actors had moved on the set—and it was too slow! Leaving the viewer aside for the moment, the first to suffer were the theatrical directors who had been imported to Hollywood on the not unreasonable assumption that they knew a great deal more about dialogue than did the movie directors. What they didn't know was film pace and screen acting, and in a remarkably short time the stage directors, with a few notable exceptions, were back in New York and most of the film directors were back in the saddle.

It seems that what plays perfectly on the set is often less than perfect when transferred to the screen. But why? The question has stumped many a filmmaker, and continues to do so. Why is a satisfactory "real" pace often slow and boring on film? Is it because the viewer is more intimately involved with the film's characters and therefore more concerned with being than with observing; that it is psychologically easier to entertain the actions and ideas of nonthreatening images than those of unpredictable real people, thus

*Take it from the horse's mouth. From 1923 to early 1929 the writer was a boy projectionist at Paramount Studio. Sound features arrived in 1928.

Shortening time. A series from Greystoke, The Legend of Tarzan *depicting the growth of the title character. Photograph courtesy of Warner Brothers, Inc.*

reducing the time needed for understanding and acceptance? Is it simply that the ever-searching camera enables the viewer to see, instantaneously and thoroughly, the full scope of the players' reactions? Or is it a combination of these and any number of other considerations that alters his perception of the scene's movement through time? Whatever. The empirical truth is that as a collaborator with shadows the viewer senses meanings more intuitively, understands dialogue more easily, and reacts to both more quickly.

This difference in sensibility which in the "silents" was accommodated by manipulation of the delivery system must now be managed by delivery, period. Creative film directors have found subtle and indirect ways of eliciting a livelier and more vital pace from their actors while avoiding the undesirable appearance of "rushing," though theatrical scripts and method techniques sometimes make their work in this area very difficult.

A single example will serve to illustrate this particular aspect of subjective time.*

We were "looping" a scene from *The Young Lions*. Monty Clift listened to a tape of himself speaking some lines. He looked puzzled and turned to me.

"That's not me," he said, ungrammatically and inaccurately.

"Of course it is," I assured him.

"It can't be," he protested. "I've never spoken that swiftly in my life."

I asked the projectionist to run the tape in sync with the matching picture. He did so, and Clift was finally convinced, though he continued to shake his head throughout the looping session.

A further refinement of this adjustment in pace is dictated by the *development* of the story. The film's opening, for instance, would naturally be presented at normal or near normal speed even if it is an action sequence since the viewer has not yet shed the effects of the real world. He is not yet acquainted with the film's characters nor is he aware of possible story trends or situations. But as the film progresses so does the viewer's familiarity with its style, its message, and its people. This growing engagement enhances his ability to think with the film's characters, even to second-guess them if the

*From Edward Dmytryk, *On Filmmaking* (Stoneham, MA: Focal Press, 1986).

filmmaker does not successively speed up his pace to match the viewer's increased awareness.

The freedom to manipulate time was recognized during the earliest filming experiments; awareness of the *need* to manipulate time followed soon after. The Keystone comedies are full of examples of obvious undercranking. Most of these are seen in comic chases. But a close examination discloses instances of very sophisticated time management which are subtle enough to escape notice in casual viewing.

The desirability of approximating widely varying aspects of subjective time has been recognized by workers in all fields of film and television, including commercials, sports broadcasting, and a host of other genre. The instant replay of an exceptional bit of football action is nothing more than an instant "flashback," a much maligned but indispensable story device long employed on the screen, and the use of slow motion or stop-frame (which freezes time) to clarify a complex move or play is also common. But there is a world of difference between fiddling with time to satisfy a viewer's desire to "see it again," and the manipulation of the rhythm, pace, and direction of time's passage to effect a vital dramatic design.

Undercranking to achieve speed is common in almost every kind of action or chase picture and, with the exception of "worst case" examples, it is rarely obvious. When "real" time and talent permit, there will even be planned variations (usually in increase of speed) as excitement builds and the viewer is pulled deeper into the film. And lest a sudden return from "fast" to "normal" result in visual confusion, the finish of such a sequence requires special treatment in bringing the viewer back to earth. The most common ploy is the use of a frozen reaction of shock—to a crash, for instance—which brings both action and time to a dead stop. When movement picks up again, time can resume a neutral pace while the viewer unconsciously re-enters the more or less normal world, both chronologically and emotionally. (It should be mentioned that the proper choice of lens and camera position is an indispensable concomitant of every time-managing technique.)

Another stratagem for imperceptibly boosting a film's pace is to curtail the time normally taken to accomplish some familiar action. This does not refer to the so-called "time lapse," a frequently used editing technique which eliminates minutes, days, weeks, or even years of irrelevant material; what is referred to here is the truncation of time to bring about agreement between an action as it will appear on the screen and the viewer's routine perception of that action. An example will clarify the concept.

It has long been known that the normal man or woman is incapable of estimating the passing or duration of time with any accuracy, and the longer the time span guessed at, the greater the degree of error. The filmmaker must learn to use such human inconsistencies to his advantage. Having had occasion to shoot a number of fight sequences—ring battles consisting of a number of precisely measured three minute rounds—I have found it nearly impossible to dramatically sustain more than ninety seconds of action unless a good deal of stalling (resting while clinching, or pointless footwork) was included. Certainly, stalling, clinching, and fancy "dancing" is largely what most fights consist of, but such action, though acceptable in comedy, is not exactly spellbinding. The viewer expects a constant flow of meaningful action, and about ninety seconds of choreographed boxing, with the real but boring stalling eliminated, will usually be accepted by the viewer as a full three minute round. That much playing time will satisfy the viewer, who has no idea he is being short-changed, and will avoid depressing those who find no special delight in the sight of two men pummelling each other into insensibility, even if the plot demands such action.

Whether achieved by mechanical means or by increased speed of playing, the imposition of an accelerated pace is, or should be, quite common in filmmaking. The opposite effect, an extension of time, is much more difficult to realize and is not often at home on the screen. *Slow motion* is accomplished *only* by overcranking and is immediately recognized as an aberration of reality. A further drawback is that it plays against the viewer's increasing alertness which accompanies the film's development. But, as usual, there are interesting and valuable exceptions in which obvious aberration of reality is apropos.

There are at least two situations in which slow motion can be used, sometimes to brilliant effect. The first, and by far the most common, is some variation of the dream or nightmare sequence, usually seen as a montage. The viewer instinctively accepts dreams as upheavals of the subconscious mind, eruptions of suppressed material in which there seems to be no logic and little, if any, continuity in action, personnel, costumes, locations, or time. But he knows that the filmmaker does have a rational resolution in mind, and he

The author rehearsing a fight scene with Robert Ryan for Behind the
Rising Sun. *Photograph courtesy of R.K.O. Radio Pictures, Inc.*

accepts the dream's pictorial realization as a metaphor which applies to some aspect of the ongoing "real" situation and, as all metaphors should, helps to clarify it.

The problem with this aspect of filmmaking is the extreme difficulty of capturing the sudden and subtle changes which shape most dreams with an instrument devised, as Kracauer put it, for the "redemption of reality." But when that instrument is properly mastered and bent to the task such montages are pure cinema, pure imagery, which serve not only to illuminate a character's hidden motivations or misconstrued emotions, but also to greatly shorten the time which would otherwise be needed if the job were done in more pedestrian, and certainly more verbal, terms.

One of the finest examples of this technique is seen in a short film of Ambrose Bierce's "An Occurrence at Owl Creek Bridge" made by Enrico in the early 1960s. The dream, which in reality would last no more than a few seconds, takes up the greater part of Bierce's story and of the 26 minute film version. In it the protagonist, who is being hanged as a spy, imagines (or dreams) that he escapes the noose and, eluding his pursuers, he makes his way toward the safety of his home. As he comes within sight of his haven and his anxious but overjoyed sweetheart, his dream approaches its end; he runs the last hundred yards hopefully, but the length of his strides diminish and become more labored with each step, until he is literally and frantically running in place. Here the telescopic lens, which shows almost no forward progress, and slow motion, which accentuates the laboriousness of his frenzied efforts, are masterfully employed to dramatize the failing mental struggle of the man who, in reality, is in the last seconds of death by hanging.

As far as I am aware, the second example of slow motion use is a one-of-a-kind examination of a shopworn action situation. The two final segments of a routine TV series, Hugh O'Brian's "Wyatt Earp," were devoted to scrutinizing the 1881 gunfight at the O.K. Corral in Tombstone, Arizona. This notorious incident, in which Wyatt Earp, his two brothers and his friend, "Doc" Holliday, faced a group of trouble-making cowboys (usually but erroneously called "The Clantons") has been either factually or fictionally used in dozens of Westerns. But the sequence generally plays for the suspense leading up to the confrontation. Like most climaxes, the fight itself does not build emotion—it releases it. The historic gunfight lasted perhaps 20 or 30 seconds. Between 40 and 50 shots were fired at extremely close range. Three cowboys were killed, and Earp's two brothers were wounded.

The actual fight, in any version, is over so quickly that nothing but the sound of the barrage, and a few quick clips of barely recognizable men, firing, falling, or fleeing, culminate the sequence. But someone in O'Brian's company decided that a frame by frame depiction of the gun battle might be more interesting than a confusing fire fight—and so it was.

During the course of some 20 to 30 minutes every move of every man, every shot from every gun, every hit and every fall, every attempt to escape, was carefully recorded. And since it is a dictum of Western folklore that a fast draw is too rapid for the human eye to follow (and it really is*) all the action was shot in slow motion. Because the treatment of the sequence was an examination rather than a dramatization, this escape from "temporal reality" was accepted as a necessary part of the presentation, and obviously meant for the viewer's enlightenment. And enlightenment, when skillfully presented, can also be entertainment.

A still different kind of time delay is sometimes needed to maintain the clarity of action or plot. All viewers are acquainted with the "in-the-meantime" convention, but a related and more complex situation requires an arbitrary postponement of one action sequence, which in real time takes place in concurrence with another, in order to maintain continuity of both. An ideal example of this technique is seen in that very fine production, *Witness*.

At the film's action climax a formidably armed trio of antagonists have cornered the unarmed protagonist in an Amish barn. Since the Amish religious belief strictly forbids any form of violence, he is completely on his own. The three heavies separate to hunt down their prey but since the farm complex is small, so is the separation, and the three men are always within each other's call. In order to avoid confusion the film shows only one confrontation at a time but, to make the prolonged and arbitrary separation believable, the editor resorted to a simple but clever expedient.

As the first round of action approaches its climax, and at just about the time the viewer might begin to wonder at the absence of

---

*It may be of interest to some film scholars to know that when Henry Fonda played a fictionalized version of Wyatt Earp in *Warlock*, his draw was so fast it had to be slowed down to make it perceptible.

the first heavy's buddies, a cut shows a second heavy making his way toward the noise of the conflict. But too late, of course. There are no surprises or obvious contradictions when the first heavy is promptly disposed of and the second appears on the scene to continue the action. The same expedient shows us the third heavy just before the second conflict is resolved, and the final sequence becomes a simple face-off which ends in a surprise denouement. During this entire portion of the film, adroit time management and the principle of isolation has enabled the film's makers to present a thoroughly understandable story and action line.

Clearly in this situation time could not have conformed to the needs of the sequence without the editor's freedom to manipulate it. It must also be clear that it is difficult to discuss the problems or the nature of film without specific or implied reference to editing or the cutting room. The reason is obvious; it is impossible to make a motion picture without an editor, and almost everybody knows that except, strangely, a few theorists and a considerable body of filmmakers who "will not see."

A film editor at work—the author at the cutting bench editing the
Mae West film, Belle of the Nineties, *for Paramount Pictures.*

14

The Force of Filmic Reality

A negative characteristic of the narrative film is simply this: It lacks the power of arbitrary recall or of impromptu timing. The viewer cannot stop the film at his pleasure to contemplate a reaction or savor a line, he cannot re-examine some earlier dialogue to clarify an ill-defined point. Nor can an actor time his performance by waiting out a possible laugh or gauge his audience's response to a vital transition, since such responses will undoubtedly vary from neighborhood to neighborhood.

Neither the writer, the director, nor the actor can with complete consistency anticipate the speed of the viewers' reactions or the extent of their involvement—so that leaves it up to the cutter. Only the film editor has some limited freedom to regulate, or change completely, the film's timing to accommodate the filmmaker and the viewer.

The classification, *Film Editor*, is unquestionably the most ambiguous credit on the film crew list. (Only the writing credit *Additional dialogue by* can compete.) The creative editor may be the director, the producer, or, in the studios, possibly an executive editorial "expert," but, at least on the few major films made each year, the director enjoys dominant control. However, with the exception of that handful of directors (all ex-cutters) who insist on handling their own films, the credited film editor does perform all the technical work required to execute the director's wishes. A few of these under-acclaimed artists have, and sometimes are allowed to exhibit, creative editing talents which far surpass those of the directors they may work for (though I have never heard a director, no matter how dim his talent, admit that he is *not* one of the world's greatest film editors). Unfortunately, only on rare occasions does one of these talented editors have final or full control.

Henry Moore wrote, "It is, of course, the vision behind a work that matters most, not the material." True. But unless the artist who furnishes that material edits his own film there can be more than one vision involved—or at the least, one vision which is not strictly pursued to the very end and another, more or less skewed, which is.

Where there is a "king" there is often an ambitious "pretender," and although the editor is usually the "director's man," he, too, has his own ideas and his own ego.* Given the opportunity and the cunning, the strong-minded creative cutter can, here and there, superimpose his own vision over that of the director. The change may not be great or obvious, and if a weak director is involved, it may be a blessing. But it can also be a curse. Many a final cut falls short of what it might have been, but because the director has no hands-on knowledge of editing, he is none the wiser. He will shake his head and wonder where he went wrong. Lucky is the non-cutting director who finds and clamps on to a top editor whose main desire is to see that the director's film fulfills its potential, for there have been more films technically and artistically damaged in the cutting room than have been saved there.

So now we come to the solid trunk of the film family tree. In the beginning, a flexible base carrying a film emulsion and the intermittent sprocket made moving pictures a reality, but moving pictures are not the same as "Film" or "The Cinema," which reaches, at least occasionally, toward "Art." And there was little possibility for art in narrative films before editing made aesthetic construction conceivable.

Until the comparatively recent "renunciation of montage" by the sequence shot minimalists, there was broad agreement on the fundamental value of film editing. The great Russian director Pudovkin said, "Editing is the force of filmic reality." Ernest Lindgren's state-

---

*The masculine gender is used here (as in most other sections) for simplicity in writing. In fact, editing is a field that has always been wide open to women (which cannot be said of most film crafts), and their skills have always been quite unchauvinistically regarded.

ment is even stronger: "The development of film technique has been primarily the development of film editing." Much more recently, Kracauer wrote, "Of all the technical properties of film, the most general and indispensable is editing." And even Bazin, who tried, without much success, to change the situation, originally agreed. "Editing," he wrote, "had once been the very stuff of the cinema." Perhaps his use of the past tense explains why the cinema is now so rarely good "stuff."

Almost everyone knows that a movie is a continuum, a nexus, consisting of innumerable pieces of film which those involved in its manufacture call *shots*, or *cuts*. These shots, some whose few frames make only a fleeting subliminal impression, others long enough to seem interminable, are cut and joined together by the film editor. But *how* does he join them together? What are his rules, his reasons, and his purposes? To be sure, he has the screenplay as a loose guide, he has the director's casual, or precise and voluminous, suggestions, usually made during the running of the daily rushes, and he has his own ideas which are fed and modified by the input from the first two sources. But obviously, though not necessarily, there must be a great deal more to it than that.

As early as 1929, hypotheses for editing films accompanied by sound were being formulated, tested, and dropped. I remember sitting in a darkened projection room just after the running of a newly cut sequence, listening to an artsy-craftsy film editor expound his theory of cutting sound films.

"If I can just lean back in my seat with my eyes closed and hear the splices go by," here he snapped his fingers rhythmically, "blip, blip, blip—I'll know the film has been perfectly cut."

It seems strange that in this modern period, when the sequence shot is scrambling for a place in the film sun, the cult of the "short cut" should be as healthy as it is. This "theory" holds that any cut held too long is a bore. Of course, at times that could be true, but the *quality* of the shot, Moore's "vision," is not included in the formula. Cutters of this elementary school are talking about cuts lasting four or five seconds, sometimes a little more. Many films of recent years have been "chopped to death," with shots abandoned before the completion of important reactions, and cuts dropped before key arguments have been fully made. Cutters brag of setting new records for the number of cuts in a reel. The concept behind this style (?) seems to be that the average viewer is interested mainly in the composition of the shot and only peripherally in its substance, that he has a short attention span and must frequently be given new pictures to look at or boredom will set in. It is most fortunate that not a single one of the world's top film editors subscribes to this aberration.

There are a number of aspects of filmmaking which fall under the jurisdiction of the film editor. First, to a greater extent than is realized by many directors, he is the author of a large portion of the film's syntax. (Think about that!) Syntax treats of the proper order of message-bearing units (images or lines of dialogue) to render the narrative more understandable (and more interesting). It is, of course, a necessary basis for all communication, and it is amenable to a great deal of manipulation, manipulation which, in literature, for instance, accounts for the differences between a Hemingway, a Melville, and a Faulkner. But film syntax is capable of even greater variations than literature, since its "words" and "rules of grammar" are not nearly as well defined or codified.

Second, Eisenstein has written, "Two film pieces of any kind, placed together, inevitably combine into a new concept, a new quality, arising out of the juxtaposition." If that statement is true—and though a good many people fear it, not too many will question it—the editor carries a heavy load. Since it is he who is responsible for at least the mechanical aspect of every juxtaposition, each of which delivers "a new concept, a new quality," he must be very careful that the juxtaposition he chooses is the most favorable, the most eloquent, in developing the dramatic elements he works with, and the most accurately made, since every frame added or subtracted from the juxtaposition alters its values. (I refer here to the creative editor who takes his responsibilities seriously. Many cutters simply put selected shots together and let it go at that. But who said the world is perfect?)

Eisenstein's pronouncement does not say that every juxtaposition of two pieces of film is profitable, only new—but the new concept might be deceptive, the new quality regressive, and the editor is faced with a decision. While each piece of film has its own meaning, its juxtaposition to another piece of film alters, extends, contradicts, diminishes, or enlarges that meaning. The result is the formation of a new or, more commonly, a modified interpretation of the cut. Since the viewer assumes that a cut, or a related combination of cuts, carries a message, the editor must be certain that each new juxtaposition carries the desired message, not an accidental one. (In editing, serendipity is rare.)

To be sure, the greatest part of such a burden should have been eliminated during production, but not all directors are maestros, and awkward conjunctions, unanticipated on the set, are not unusual. Careless shooting is too common on many films. And for a number of legitimate reasons films are always shot out of sequence; no matter how conscientiously the director, the actors, and the script supervisor monitor the scenes, slipups in the consistency of mood and tempo are bound to surface. Even masterpieces will suffer a glitch or two.* Which is why a good editor is rarely unemployed. The final tailoring, the smoothing out of juxtapositions whose junctions are "out of sync" is in his hands. The truth is that the most creative filmmakers are inclined to gamble, to guess, and if an occasional guess is wide of the mark, the first corrective move is a try for a miracle in the cutting room.

Juxtaposition is only one of the editor's many problems. The selection of the shots which will be joined together, one after another, until the sequence at work is completed, comes first. The logical questions then are, What are the requirements for choosing a shot? and Why is one cut preferred over another for any part of a scene?

It has already been noted (page 129) that the director may have indicated his choice of angle and performance (good actors rarely play a scene exactly the same in different takes or set-ups). And unless, for some unexpected reason such as impossible matching, the director's choice will not "work," the editor will follow his wishes if they have been clearly articulated. As for the rest, which is usually a major portion of the day's rushes, the editor is the final decision maker, at least through the first cut. And though some might think that the scenes as shot are so specific as to negate the possibility of playing around, the everyday experience of the cutting room demolishes that notion.

Ideally, the editor's perceptions of a scene's values should match those of the director. But at times, if he is very clever and sufficiently alert, he may come up with more because, every now and then, close scrutiny of a newly edited sequence will uncover an unexpected

*However, when a thirty-five million dollar production lives in the cutting room for a year or more, one may surmise the shooting was, to say the least, confused.

nuance, or suggest an interesting twist in the scene's development. Bonuses will arise from a surprisingly original performance, or from a fresh slant on characters that surfaces only after juxtapositions have been made. But it all starts before a single cut has been considered.

After the first or second running of the dailies, many cutters take the film directly to the Moviola or the flat-bed, but when a complex sequence is involved this is not the procedure of choice. Since the images on either of these machines are too small to deliver an exact focus of attention, the editor with a disposition for thorough analysis may look at his rushes repeatedly to fix the significant details— transitions, key reactions, performance, and occasionally pictorial effects—firmly in his mind. And though he uses the Moviola to locate the exact frames for his cut, he will have made his cutting decisions on the basis of what he saw on the "big screen," because the messages seen there are the only ones to be seen later by the viewer.

From the very first cut, which may be a long shot to establish the milieu or a close-up to introduce a character, the editor selects cuts which, at least in his opinion, say what the sequence at any particular point should be saying. There will be occasional compromises as when, for example, a performance in a selected close-up is not as effective as that given in a tight group. Here, the editor must decide whether more is to be gained from the closeness, the search of the actor's eyes, or from the slightly superior playing in a somewhat longer shot.

Although a few set-ups may be shot for "protection," most will have been created for a specific purpose, and each will be sought out and used with that purpose in mind. So the next question is: How long should each shot run? The answer is remarkably simple: It should run as long as it is the best cut for delivering the immediate message, whether that is six frames or one thousand feet—but not *one frame* longer. When another shot says it better it should *instantly* replace its predecessor.

However, within that general rule there are necessary variations; the length of the cut will also depend on the nature and character of the information it must convey. Every juxtaposition may "combine into a new concept, a new quality," but that is hardly a good reason for rushing on to the next splice. As a rule the viewer is not aware of a well-made juxtaposition, and the screen image, whether beautifully or badly shot, tells him a great deal more. Now we're talking a universal language. The image delivers its message, good or bad, pleasant or repulsive, touching or insensitive, and the viewer's acceptance of the cut's implications is controlled to a large extent by the time allowed for its contemplation. In other words, the strength and the depth of a character's reaction to a spoken thought or a physical deed is determined by the nature of the stimulus, and the time required for understanding that stimulus will vary with each viewer's intelligence and quickness of wit. Most of the time the difference will be very slight and can be ignored. But if the stimulus is complex or profound the reaction cut should be long enough to ensure the decipherment of the stimulus and the resultant reaction. Now the time difference in viewer understanding may be measurable, and since the editor can't please everyone he will time his cut to accommodate the slower mind. The increase in length will probably not be too great but in extreme cases, if the filmmaker has been lucky or wise enough to cast the best, the editor will have little to concern him. A Jack Lemmon or a Diane Keaton can keep an extended cut alive for even the quickest mind.

In short, cutting demands are imposed by the story, the attendant situation, the complexity of the particular sequence, the characters, the mood, and the pace, not by some esoteric theory. A bravura sequence shot such as the opening of *A Touch of Evil*, perfectly serves its chief purpose, that of showing off the director's conceptual skill (we have all succumbed to the temptation, and for the same reason) but its construction and execution is really a triumph for the crew, especially the cameraman and the grips. As a shot of substance it is probably 5 on a scale of 10. Close analysis uncovers several spots which would have been better served by more specific set-ups.

Even in the field of routine films where plagiarism is the rule, no two sequences are quite alike; they are as different as snowflakes and fingerprints. And no specific equation can be devised to handle all the variables. Only a general rule can approach validity, and its very generality invites increased diversity. Although two editors would undoubtedly differ somewhat in their value judgments when cutting identical material, they would probably agree within a frame or two where each cut should begin and where it should end, regardless of the differences in editing. The same could never be said for the average director and his editor. Here is where an instinct for optimum pacing asserts itself. A few frames too many at the start of a cut, a few frames left dangling at the end, multiplied by the number of juxtapositions in the film, will lead to a hesitant and lagging pace. Such a lapse may last only a fraction of a second, and even a good director will find it difficult to isolate the problem, but only a few editors have an instinct for the precise frame which does not overstay its welcome for even a fifth of a second. And only one film in a hundred shows the mark of such sensitive editing.*

Excluding the many pure action movies, most films made today are essentially theatrical; that is, dialogue carries the day. The best are a mixture of cinema and theater; some sequences are all dialogue, others are largely action, or image, and this imposes a duality of techniques on the editor. Where dialogue dominates, it should be respected and given its full due; but when action or image prevails, words, like music, become underscoring. In short, a sequence should get the treatment that will exploit its special qualities, but it should be understood that the two techniques are quite different.

Until now the words "cutting" and "editing" have been used interchangeably, as they still are in Hollywood, but such ambiguity diminishes our ability to discuss the craft intelligently. In my lexicon the two words are by no means synonymous, except in a very casual sense, and the fact that they have been so used probably helps to explain the decline in regard and respect for the art and the artist. All editors can "cut," but only a comparitively few "cutters" can edit. Advanced editing is an intuitive skill and, as is true in all arts and crafts, it is exceedingly rare.

It is a cliché in the field of education that "talent can't be taught," and since editing is a gift rather than a mechanical routine, it is difficult to illustrate its advantages, but two simple examples will at least hint at the concept. The first is that of the football game in Chapter 3, where it is used in a different context. But if the change in routine, as suggested, takes place during the cutting process rather than in the shooting, it is called creative editing. The second example is slightly different. Let us imagine that a sequence showing the start of an important airplane flight has been written and shot in the following manner: First, a long shot of a busy airport with planes taking off and landing, then a closer shot of a particular plane at its

---

*For a detailed technical exposition of this aspect of film editing, see Edward Dmytryk, *On Film Editing* (Stoneham, MA: Focal Press, 1984), Chapter 9.

terminal. A montage of cuts shows passengers getting out of taxis or limousines, crowding the ticket lines, depositing luggage, negotiating the electronic frisking gates, boarding the plane, and finally settling in their seats. About this time a nervous baggage handler approaches the plane, looks about carefully, then quickly opens a luggage door and shoves in a small valise.

Now, of course, the viewer feels some suspense, but how late in the proceedings! Here, the preliminaries, even if nicely edited, have been routine and dull, more documentary than narrative in style. Characters have been introduced in some of the montage cuts, but even if they are "names" they are of little immediate importance.

However, a smart editor can realign the material to show the stowage of the suspicious bag *before* the boarding montage develops. Now the passengers' very lack of apprehension has an aura of suspense; the introduced characters as well as the background people are all potential victims and therefore worth the special attention of the viewer, who is aware that he may be looking at dead men and women.

Such rearrangements of cuts, usually far more complicated than these beginners' examples, seem quite simple and logical. Why aren't they written that way? Why wasn't *Heaven's Gate* a smash success? Hindsight is always easy, but such misconstructions in the writing and the shooting are by no means infrequent, and the possibility of an improved scene are often overlooked. Still, creative editing (which can stem from any one of a half-dozen minds) has improved a thousand sequences.

Finally we come to one of the most difficult and sensitive areas of good film editing—the film as a singularity. For the film as a whole, the extension of the properly timed cut is the perfectly smooth flow. In this context, "smoothness" is quite separate from the film's level of tension, of action, of violence; it underlies all of these but deals only with the inexorable progress of time—the flow of life. Although, in the process of shooting, scenes have been studied, started and stopped a thousand times and more, the edited result should show not the slightest evidence of this, only a subconscious impression of continued forward movement. Temporal reality is unimportant, but chronological *direction* is an imperative. The river of time may have eddies, rapids, and limpid pools, but it must never stop, hiccup, or flow backwards. And the viewer must feel, if he bothers to think about it, that every character in the film, no matter how insignificant, continues to live and, unless he is killed or dies in the film, will continue to live even when he is offscreen. Life does not stop with the cut.

15

About a Forgotten Art

"I want to share this with you." During the last two decades this expression of the "Me" generation's idea of unselfishness has captured the young and the old alike. Such a concept destroys not only marriages but artists as well.

In Southern California, an elderly TV weatherman often "shares" his meteorological information. Why doesn't he give it away? If he is reluctant to bestow that which is not really his to begin with, he might at least "pass it on" with apologies to the United States Weather Bureau. But no, he *shares* it, even though giving would cost him not a pennyworth in money or self-esteem, and would relieve him of a very unreliable responsibility. The ancient desert patriarchs knew the score. "It is more blessed to give," they said—to *give*, not share. "It is more blessed to give than to receive."

This book's orientation has been toward the receiver, the viewer, because although giving is never wholly altruistic, the chief goal of any artist should be to give his work to the people. To create for one's own glory is risky business, since no artist knows for sure the aesthetic worth of his work while it is in progress. But if people accept his efforts, they will crown him with more garlands than his brow can decently wear. On the other hand, the man who asks his crew to perform the most difficult feats purely for his own credit may dazzle thousands but he is, at least at the moment, an arty pretender rather than an honest artist. Filmmaking is not just an opportunity to make an obvious display of one's creative brilliance;

it is an effort to communicate effectively, to reach and touch another. I would gladly trade the most ingenious ten minute "sequence shot" for a 27-second close-up of Ingrid Bergman looking longingly at the man she loves. The less gifted filmmakers offer their viewers technique or catharsis and escape; the artists help them to learn how to live.

"Poetry comes with anger, hunger, and dismay," said Christopher Morley, and Donald Culross Peattie affirmed what the world of artists had long accepted as a cliché. "A poet should always be hungry," he wrote. For obvious reasons narrative film is rarely poetic; normally it is a mass medium which aims its product at a mass audience. It is freely charged and reluctantly admitted that the resulting product is overwhelmingly mediocre, and a mediocre product suggests mediocre crafting and craftspeople. So it is little short of miraculous that in a medium dedicated to the ordinary, a few filmmakers have been able to extrude gold out of blocks of utterly common clay, to create "poetry" through narrative film.

Today the miracle is less often in evidence; most filmmakers, like their product, now spring from the middle class, and their only spur is the possibility of money and fame. As indicated by Morley and Peattie, those are not the incentives that have traditionally propelled anyone toward art. Perhaps a few women and minority members, predominantly blacks and Latinos who are beginning to edge their way into the filmmaking community against fairly heavy odds, have been gouged by the rowels of hunger and anger, but film recruits are largely fat and sassy college graduates. Few, if any, have felt the wolf gnawing at their bellies, and most reserve their ire for those who deny them their luxuries or try to prod them to higher standards of excellence.

The "democratic" contention, spread by parents and educators, that all persons are born with a "gift" that can be uncorked by training and practice is, of course, the sheerest of nonsense. Only an exceedingly frivolous definition of "talent" could possibly support that point of view. But train them we do, with the result that competent "visual" engineers are turned out by the hundreds.

"We have a lot of good engineers," writes Allan Bloom in *The Closing of the American Mind*, "but very few good artists." Which is fair enough, and by no means unusual. But while developing its engineers, society has lost sight of its obligation to discover and nourish (or starve) the artists. To be sure, engineering, of any kind,

is an honorable and useful profession, but engineers and artists walk
separate ways. No bridge can be built without recourse to a mul-
titude of formulas and equations, but no prescriptions exist which
guarantee the construction of a good film, and the director who
thinks he has found them by slavishly copying the work of his betters
is not a filmmaker but a hack.

So here we are, at an impasse, high on the horns of a dilemma,
stuck with few rules and even fewer rule breakers, which is either
a catastrophe or the dawn of a new era. While the situation is an
aesthetic disaster for the visual engineer, it can be a breakthrough
for the artist; the fewer the rules the freer the creative mind, if one
can find it. In the early 1950s only a few people were aware that the
adoption of technical innovations such as Cinemascope and Cinerama, and the development of advanced methods of recording and
delivering sound were, at least in part, the industry's desperate attempts to fill the growing gap left by the deterioration of film quality.
Well, the more things change. . . . Today the air is still filled with
news of technical wonders; tiny chips which can record an entire
film are just around the corner, and the miracle of digital sound is
already with us. There is much talk of better *looking* films, of better
sounding films, but none about *better* films.

However, despite the arresting of film's aesthetic growth a few
stunted but still green twigs continue to survive, an indication,
perhaps, of a long delayed regeneration. If that is so, there is no
question that motivated creators must soon be found. But how?
What are the substitutes for the traditional motivations of hunger
and anger so often cited by the poets? Wishing won't bring them
back and, in any case, the wisher would be attacked as a regressive
and sadistic animal.

And where? The discouraging truth is that we are forced to look
for creative minds in a society that breeds far more criminals than
creators, that stultifies rather than inspires, a society that cheerily
substitutes a Pee Wee Herman for a Groucho Marx or a Rodney
Dangerfield for a Charlie Chaplin, and accepts the Roman numerals
II, III, IV, and even V, as completely satisfactory stand-ins for originality and innovation. Even more depressing is the fact that we
have hatched a youthful community far more interested in special-
ization, in the development of techniques and formulas, a generation
that does not recognize that the understanding of substance, "soul,"
and the human condition, without which no film can really breathe,

can be acquired only through a continuing study of intellectual disciplines that are in no way related to filmmaking techniques.

Although substance, "soul," and a concern for the human condition are the main ingredients of any good film, they can be blended only by those who have a special talent, a "gift," for filmmaking, and that talent can no more be taught in a book than in a classroom. All that can be explored here is the nature of any possible incentive that will prod the gifted few to come to the aid of a nearly extinct art form, and a major irritant comes immediately to mind: We are Earth's most bellicose creatures and even the well-fed can rarely resist a challenge, if it is recognized and admitted.

It would appear to be obvious that the "dialogue" film, trailing the long heritage of the theater behind it, has little capacity to challenge or to develop. It is going nowhere. Except for the presence of some theatrical acting and the absence of a good deal of lurid language, the good dialogue film of a half-century ago—say Capra's, *It Happened One Night* (1934)—is still at least the equal of the best of our time. That should be no surprise. No playwright in almost four hundred years has been accepted as a peer of Ann Hathaway's husband, so the road paved with spoken words can be viewed as leading to no new or higher ground. The only challenge is to do what everyone else is doing but to do it better—which should hardly be enough. But moving images are a recent invention, an even more recent art, and Griffith, Eisenstein, von Stroheim, and Murnau did not exhaust the medium's possibilities—not by any means. Why, then, don't young directors accept the challenge to make films with developmental potential, to feel the excitement that only the birth of something new can bring? There are three basic reasons for this state of affairs.

The first, quite simply, is that few film students or young directors are aware that an alternative exists. The second is that it is *much easier*, both technically and mentally, to make and to cut a dialogue film. To be sure, it requires patience, perseverance, and an inventive mind to delete, reorganize, emend, and generally try to improve the average talking scene (a good writer makes it all much easier), but that is not nearly as difficult as staring into space or pacing the floor for hours trying to find an original way to create screen metaphor out of "To be or not to be." Of course, it has never been done. But has anyone ever tried?

The third reason is that the environment and the conditioning needed to reanimate the art are nearly nonexistent; neither is readily available.

[Author's note: Vestiges of montage technique still exist, as do a few directors who can still communicate through imagery. One has only to see a Fellini, a Lean, or a Forman film to sense how much thought and work was involved in the creation of visual storytelling. And two recent films, *Greystoke*, and *The Man from Snowy River*, stand out as excellent examples of balanced correlation of image and dialogue. Extremely long sections of each film are played in silent action (largely nonviolent and often idyllic) which is musically underscored. The Australian film is nearly ideal in its selective "sandwiching" of talk and picture. Furthermore, it is only fair to say that "talking" scenes occasionally present opportunities for creative pictorial treatment, though these opportunities are rarely realized since the modern mind-set is predominantly toward dialogue. But as far back as *Mr. Smith Goes to Washington* (1938), Jimmy Stewart's last ditch filibuster was perfect material for the treatment of dialogue as underscoring—for the creative visual presentation of the lone, "little" man desperately struggling for his principles which, at that point, needed no further elucidation. Here, body and face language were dramatically more important than the words. In a sequence of this sort, what is seen is much more to the point than what is heard and, once in a while, imaginative treatment might turn it into a cinematic classic.]

"The development of film technique has been primarily the development of film editing." This statement is worth repeating, for with that observation Ernest Lindgren accurately assessed the condition of the narrative film from his day to the present. Indeed, film editing has not changed a whit in the last fifty years or more, and neither have films. That is the negative side of "Lindgren's Law." The positive side is that a developmental revolution in the art of editing should inevitably have beneficial repercussions on the style and quality of the narrative film as well.

The editing of a dialogue sequence requires only basic skills. The script itself sets the direction and the emotional tone of the scene which the cutter almost always respects. Rarely can he interpolate a useful "outside" shot into the flow of the scene. It is obligatory that cuts are placed in proper sequence as written or shot, and made invisible. But a sequence of gripping suspense or, better yet, of physical and emotional movement combined as, for example, the hunt

Suspense in a close-up. A shot of Peter Lorre in Fritz Lang's M.

for the child killer in *M*, or the massacre on the Odessa steps in *Potemkin*, require not only technical skills but a burst of creative talent. Although such sequences are loosely based on a continuity, in the editorial realization the sky is the limit. The real creativity begins when the cutting gets under way, when the person editing the film starts "laying the bricks." The contributing cuts have a life of their own which the editor must discover—they can be realigned, repeated, placed in contradiction or in contrast to each other, and freely manipulated to create the maximum emotional effect.

A paraphrase of Eisenstein's dictum puts it simply: In a montage what matters is not only how the cut looks by itself, but how it looks in juxtaposition to the cuts which accompany it. That is *not* the same as cutting from a verbal or pictorial stimulus to an obligatory reaction. In the latter case the cut is *mandated* by the material; in a montage a cut is *chosen* because of its distinctive effect, alone or in association, on the viewer's emotions.*

All this is heady stuff, the "stuff" Bazin was talking about (page 129). But trying to make it something more than just a very infrequent few hours of exhilaration brings up the question: Who's to do it? Editing at this level requires a great deal of conceptual preplanning, not just a "wait till I get it in the cutting room" attitude, and editors could hardly manage it; they have neither the opportunity, the time, nor the creative freedom to mount such a broad new attack on what has become a narrow old technique. For much the same reasons the move cannot be initiated by the routine film director, and he suffers the additional drawback that his knowledge of cutting is on a par with that of a screenwriter. The impetus can come only from filmmakers who consider editing a significant part of their creative endeavors. Even a casual examination of the dissimilarities between the 1920s and the following decades shows one thing clearly; the great directors of the earlier period were all masters of montage who could, and undoubtedly did, spend much time in the cutting rooms, as do the very few who can still create cinematic film today.**

---

*In this book the word *montage* is used in the Vorkapich sense to mean a creative alignment of essentially silent cuts, not as Hollywood uses it to indicate passage of time or a bad dream.

**The months I spent running hundreds of reels of von Stroheim's *The Wedding March* as he struggled to cut it down from 126 reels to a releasable length was an irreplaceable experience. He never made it, but the longer he worked the more I learned.

As mentioned above, cinematic filmmaking demands editorial preparation.* A thorough knowledge of "hands on" editing will enable a director not only to cut the stuff he shoots but, more important, to anticipate his cutting strategy to the great benefit of his shooting. Even before he plots his staging and his set-ups the director must have a fair approximation of what kinds of juxtapositions will advance his concepts. A competent editor can cut the film he is given, and often cut it well, but he cannot undertake the planning and the shooting with creative cutting in mind. Only the director can do that—if he knows how. And that is the main problem for the future.

All directors, no matter how green, are confident that they are editing experts, that they have a "natural flair," and therefore they need not spend time learning the craft. However, in any line of work natural flair is useless until the lucky possessor has learned how to profitably apply it, and of all the crafts of filmmaking, editing is probably the most difficult to master. This is especially true of montage; a couple of semesters in a school of the cinema will hardly scratch the surface. After six months in a cutting room an apprentice will confide that editing holds no secrets from him, even though he has not yet laid scissors to film, while an experienced feature editor will confess that it was closer to ten years before he knew what editing was about, and only then did he begin to see the great range of possibilities in his craft.

It is, of course, extremely irrational to expect young students and neophyte directors to make a headlong rush for the cutting benches, but perhaps not too much to hope that at least the odd talent, here or there, wisely in less of a hurry to "make it" (an attitude that has sidetracked more hopefuls than it has "made") and more dedicated to future results, will take a long step backwards to gather up the

---

*Some directors anticipate their shooting and their editing by the use of "story boards." This is the equivalent of sketching cartoons for a painting, and tends to "set" the sequence in terms of static composition rather than dynamic movement. But I was surprised to read in an interview with Edwin Mullins in 1978 that Henry Moore said, "In my early stages I . . . made drawings . . . before making the sculpture. . . . Now I've gone away from that thinking because . . . the drawing becomes too much a key view you refer to." Even in a static art! Equally dangerous for the lazy director is the temptation to accept the sketches as the final conception so that little further developmental thought is devoted to the staging and the shooting of the story-boarded sequence.

severed threads of a prematurely abandoned craft, then take two steps into the future to start a long overdue renaissance of what once promised to be the most dynamic art the world had known.

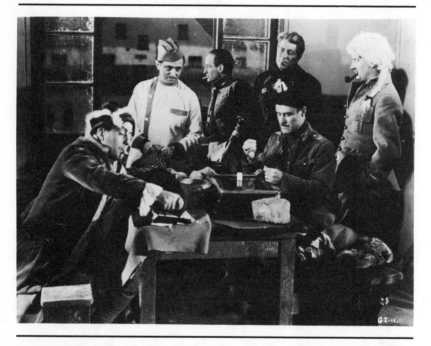

One of the best films ever made. A group shot from Jean Renoir's La
Grande Illusion.

Postscript

I have discussed filmmaking largely from a Hollywood point of view because, although I have made many films abroad, the cutting rooms and stages of Hollywood were my "film school," and the techniques and concepts I learned there are what I know best. However, in my youth those cutting rooms and stages were peopled by men like Boleslavsky, Buchowetski, von Stroheim, Murnau, Renoir, Lang, and Lubitsch, and I am perhaps more conscious than most Americans of the heavy debt we owe the great filmmakers of Europe. If I say that I still consider Renoir's *La Grande Illusion* to be the best film ever made, I think I will have said it all.

POSTSCRIPT

Filmography
of
Edward Dmytryk

THE HAWK   (Ind)   (1935)
TELEVISION SPY   (Para)   (1939)
EMERGENCY SQUAD   (Para)   (1939)
GOLDEN GLOVES   (Para)   (1939)
MYSTERY SEA RAIDER   (Para)   (1940)
HER FIRST ROMANCE   (I.E. Chadwick)   (1940)
THE DEVIL COMMANDS   (Col)   (1940)
UNDER AGE   (Col)   (1940)
SWEETHEART OF THE CAMPUS   (Col)   (1941)
THE BLONDE FROM SINGAPORE   (Col)   (1941)
SECRETS OF THE LONE WOLF   (Col)   (1941)
CONFESSIONS OF BOSTON BLACKIE   (Col)   (1941)
COUNTER-ESPIONAGE   (Col)   (1942)
SEVEN MILES FROM ALCATRAZ   (RKO)   (1942)
HITLER'S CHILDREN   (RKO)   (1943)
THE FALCON STRIKES BACK   (RKO)   (1943)
CAPTIVE WILD WOMAN   (UNIV)   (1943)
BEHIND THE RISING SUN   (RKO)   (1943)
TENDER COMRADE   (RKO)   (1943)

MURDER, MY SWEET   (RKO)   (1944)
BACK TO BATAAN   (RKO)   (1945)
CORNERED   (RKO)   (1945)
TILL THE END OF TIME   (RKO)   (1945)
SO WELL REMEMBERED   (RKO-RANK)   (1946)
CROSSFIRE   (RKO)   (1947)
THE HIDDEN ROOM   (English Ind.)   (1948)
GIVE US THIS DAY   (Eagle-Lion)   (1949)
MUTINY   (King Bros.-U.A.)   (1951)
THE SNIPER   (Kramer-Col)   (1951)
EIGHT IRON MEN   (Kramer-Col)   (1952)
THE JUGGLER   (Kramer-Col)   (1953)
THE CAINE MUTINY   (Kramer-Col)   (1953)
BROKEN LANCE   (20th-Fox)   (1954)
THE END OF THE AFFAIR   (Col)   (1954)
SOLDIER OF FORTUNE   (20th-Fox)   (1955)
THE LEFT HAND OF GOD   (20th-Fox)   (1955)
THE MOUNTAIN   (Para)   (1956)
RAINTREE COUNTY   (MGM)   (1956)
THE YOUNG LIONS   (20th-Fox)   (1957)
WARLOCK   (20th-Fox)   (1958)
THE BLUE ANGEL   (20th-Fox)   (1959)
WALK ON THE WILD SIDE   (Col)   (1961)
THE RELUCTANT SAINT   (Col)   (1961)
THE CARPETBAGGERS   (Para)   (1963)
WHERE LOVE HAS GONE   (Para)   (1964)
MIRAGE   (Univ)   (1965)
ALVAREZ KELLY   (Col)   (1966)
ANZIO   (Col)   (1967)
SHALAKO   (Cinerama)   (1968)
BLUEBEARD   (Cinerama)   (1972)
THE HUMAN FACTOR   (Bryanston)   (1975)